MATERIALS FOR ENGINEERING

MATERIALS FOR ENGINEERING

2nd Edition

J. W. Martin

Department of Materials
The University of Oxford

MANEY
publishing

FOR THE INSTITUTE OF MATERIALS

Book B0756
Second edition first published in 2002 by
Maney Publishing for
The Institute of Materials
1 Carlton House Terrace
London SW1Y 5DB

First published 1996

Maney Publishing
is the trading name of
W. S. Maney & Son Ltd
Hudson Rd
Leeds LS9 7DL
ISBN 1-902653-50-5

Typeset and Printed by Alden Group,
Great Britain

Contents

Chapter 4: Glasses and Ceramics   129

Chapter 5: Organic Polymeric Materials   155

PREFACE TO THE SECOND EDITION

Since the appearance of the first edition of the book, it has been pointed out to me that its value to the student reader would be increased if a series of related problems were included. Over 50 such problems have been devised, and they appear at the end of the text.

The opportunity has also been taken to correct a number of misprints and errors which appeared in the earlier edition. I am particularly indebted to Professor Christopher Viney of Heriot-Watt University for his assistance in this regard.

John W. Martin.
November 2000

PREFACE TO THE FIRST EDITION

This textbook represents an attempt to present a relatively brief overview of Materials Science, the anticipated readership being students of structural and mechanical engineering. It is in two sections – the first characterising engineering materials, the second considering structure/property relationships.

Emphasis is thus placed on the relationship between structure and properties of materials, starting with the concept of 'structure' at three levels – *crystal structure, microstructure, and molecular structure*. The discussion of microstructure introduces the topics of phase transformations, metallography and phase diagrams – none of which would be familiar to the intended readership.

After a section on the determination of mechanical properties, the remaining four chapters deal with the four important classes of engineering materials, namely metals, ceramics, polymers and composites. It is estimated that there are some 40 000 metallic alloys in existence, over 5 000 polymers and some 2 000 ceramic materials, so there is some justification for discussing metals and alloys at the greatest length. In that chapter, an attempt has been made to consider initially the general principles of strengthening, so that the individual families of engineering alloys can be discussed in the light of this introduction. About equal emphasis is placed on the remaining classes of materials.

The Tables of Data within the text, and the Appendices have been selected to increase the value of the book as a permanent source of reference to the readers throughout their professional life. The latter include:

- The Periodic Table of the elements
- Useful Constants
- Unit Conversion Factors
- Selected data for some elements
- A list of sources of Material Property Data, in the form of both Handbooks and Database software

The author is pleased to acknowledge the encouragement and suggestions given by the members of the University Books Subcommittee of The Institute of Materials. I am also most grateful to Professors B. Cantor and D. G. Pettifor, FRS, for the facilities they have kindly provided for me in the Oxford University Department of Materials, and to Peter Danckwerts for his efficient dealing with editorial matters.

John W. Martin.
September 1996

PART I: CHARACTERISATION OF ENGINEERING MATERIALS

INTRODUCTION

The materials available to engineers for structural applications embrace an extremely wide range of properties. We can classify them into **THREE** broad families as follows:

METALS & ALLOYS

ENGINEERING CERAMICS & GLASSES

ENGINEERING POLYMERS & ELASTOMERS

There is the further possibility that materials from two or more of these families may themselves be combined to form a **FOURTH** family, namely:

COMPOSITE MATERIALS.

It is possible to present a broad 'overview' of the properties of engineering materials by constructing a *Material Property Chart*. These show the relationship between two selected engineering properties of the above families, and Fig. 1.1, due to Ashby illustrates the *Young's modulus – density* chart for engineering materials.

Young's elastic modulus is one of the most self-evident of

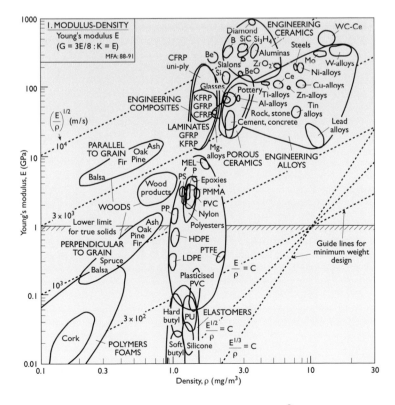

Fig. 1.1 Young's modulus, E, vs. density, ρ. (After M.F. Ashby: *Acta Metall.* (1989 **37** 1273).

material properties, reflecting as it does the stiffness of structural steel or the compliance of rubber. Because of this wide range of values, the scales of the axes in Fig. 1.1 are logarithmic, and their ranges have been chosen to include all materials from light polymeric foams to engineering alloys.

Data for a given family of material are seen to cluster together on the chart, and each family has been enclosed within an envelope in the diagram. Although each class of material has characteristic properties, these may vary *within* each class because of variations in *structure* at **three** different levels, namely the *atomic arrangement, or crystal structure,* the *microstructure,* which refers to the size and arrangement of the crystals, and the *molecular structure.* We will consider these aspects of structure in turn.

The Structure of Engineering Materials

CRYSTAL STRUCTURE

This refers to the ordering of atoms into different crystalline arrangements. It is the arrangement of these atoms – the strength and directionality of the interatomic bonds – which determines the ultimate strength of the solid. Techniques involving X-ray or electron diffraction are employed to determine crystal structures, and four types of interatomic bonding are recognised: van der Waals, covalent, ionic and metallic. The latter three 'primary' bonds are limiting cases, however, and a whole range of intermediate bonding situations also exist in solids.

The **van der Waals** force is a weak 'secondary' bond and it arises as a result of fluctuating charges in an atom. There will be additional forces if atoms or molecules have permanent dipoles as a result of the arrangement of charge inside them. In spite of their low strength, these forces can still be important in some solids, for example it is an important factor in determining the structure of many polymeric solids.

Many common polymers consist of long molecular carbon chains with strong bonds joining the atoms in the chain but with the relatively weak van der Waals bonds joining the chains to each other. Polymers with this structure are thermoplastic, i.e. they soften with increasing temperatures and are readily deformed, but on cooling they assume their original low-temperature properties but retain the shape into which they were formed.

Covalent bonding is most simply exemplified by the molecules of the non-metallic elements hydrogen, carbon, nitrogen, oxygen and fluorine. The essential feature of a covalent bond is the *sharing* of electrons between atoms, enabling them to attain the stable configuration corresponding to a filled outermost electron shell. Thus an atom with N electrons in that shell can bond with only 8–N neighbours by sharing electrons with them.

For example, when $N = 4$, as in carbon in the form of diamond, each atom is bonded equally to four neighbours at the corners of a regular tetrahedron, and the crystal consists

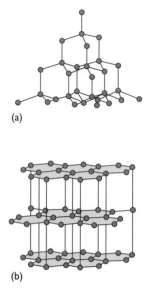

(a)

(b)

Fig. 1.2 The crystal structures of (a) diamond and (b) graphite.

of a covalent molecule, Fig. 1.2(a), and is one of the hardest materials known. In graphite only three of the four electrons form covalent bonds, so a layer structure forms, Fig. 1.2(b), and the fourth electron is free, which gives some metallic properties to this form of carbon. Graphite crystals are flat and plate-like, and they are so soft that graphite is used as a lubricant. It is clear from Fig. 1.2 that the different dispositions of the covalent bonds in space have a profound influence on the atomic arrangements and hence upon properties of the material. A third form of carbon has recently been identified in which groups of atoms group into tiny spheres or tubes. This exciting observation has not yet led to the development of a material of engineering significance, however.

The elements can be divided into two classes, *electronegative* elements (such as oxygen, sulphur and the halogens) that tend to gain a few electrons to form negatively charged ions with stable electron shells, and *electropositive* elements (such as metals) that easily dissociate into positive ions and free electrons. **Ionic bonding** consists of an electrostatic attraction between positive and negative ions. If free atoms of an electropositive element and an electronegative element are brought together, positive and negative ions will be formed which will be pulled together by electrostatic interaction until the electron clouds of the two ions start to overlap, which gives rise to a repulsive force. The ions thus adopt an equilibrium spacing at a distance apart where the attractive and repulsive forces just balance each other.

Figure 1.3 shows a diagram of the structure of a sodium chloride crystal: here each Na^+ ion is surrounded by six Cl^- ions and each Cl^- is surrounded by six Na^+ ions. Many of the physical properties of ionic crystals may be accounted for qualitatively in terms of the characteristics of the ionic bond: for example they possess low electrical conductivity at low temperatures but good ionic conductivity at high temperatures. The important ceramic materials consisting of compounds of metals with oxygen ions are largely ionically bonded (MgO, Al_2O_3, ZrO_2, etc).

Metallic bonding. About two-thirds of all elements are metals, and the distinguishing feature of metal atoms is the looseness with which their valence electrons are held. Metallic bonding is non-directional, and the electrons are more or less free to travel through the solid. The attractions between the positive ions and the electron 'gas' give the structure its coherence, Fig. 1.4. The limit to the number of atoms which can touch a particular atom is set by the amount of room available and not by how many bonds are formed. 'Close-

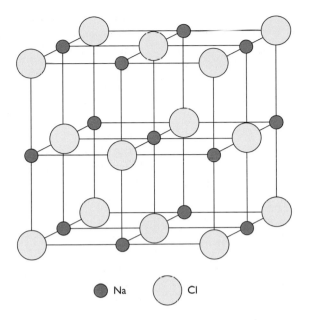

Na   Cl

Fig. 1.3 The crystal structure of sodium chloride.

packed' structures, in which each atom is touched by twelve others, are common, and they give rise to the typical high density of metals. Since each atom has a large number of neighbours, the overall cohesion is strong, so metals compare well with ionic and covalent solids as regards strength and melting point.

In general, the fewer the number of valence electrons an atom has, and the more loosely they are held, the more metallic the bonding. Such elements have high electrical and thermal conductivities because their valence electrons are so mobile. Although a satisfactory description of some of the physical properties of metals can be obtained from this 'free electron' picture, many other properties (particularly those concerned with the motion of electrons within metal crystals) have to be explained in terms of electrons as waves occupying definite quantised energy states.

As the number of valence electrons and the tightness with which they are held to the nucleus increase, they become more localised in space, increasing the covalent nature of the bonding. Group IVB of the Periodic Table (see p. 233) illustrates particularly well this competition between covalent and metallic bonding: diamond exhibits almost pure covalent bonding, silicon and germanium are more metallic, tin exists in two modifications, one mostly covalent and the other mostly metallic, and lead is mostly metallic.

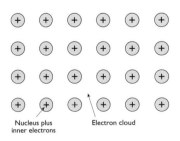

Nucleus plus inner electrons   Electron cloud

Fig. 1.4 The classical model of a metal crystal.

MICROSTRUCTURE

This refers to the size and arrangement of the crystals, and the amount and distribution of impurities in the material. These features are typically of a dimensional scale of 1–100 μm, and they determine many of the properties of metals and ceramics.

An Introduction to Phase Transformations

The transition from the liquid state to the solid state is known as 'crystallisation', and the mechanism by which the process takes place controls the microstructure of the final product. A phase transformation, such as the change from liquid to solid, occurs by a mechanism of *nucleation* of small 'seed' crystals in the liquid, which then grow by the addition of more material from the liquid. The driving force for this change can be obtained by considering the change in free energy on solidification. For example, if a liquid is undercooled by ΔT below its melting point (T_m) before it solidifies, solidification will be accompanied by a decrease in Gibbs free energy of ΔG. The Gibbs free energy of a system is defined by the equation

$$G = H - TS$$

where H is the enthalpy, T the absolute temperature and S the entropy of the system. The free energies of the liquid and solid at temperature T are given by

$$G^L = H^L - TS^L$$

$$G^S = H^S - TS^S$$

so that at temperature T, the change in free energy/unit volume upon solidification may be written:

$$\Delta G_v = \Delta H - T\Delta S \qquad (1.1)$$

where $\Delta H = H^L - H^S$ and $\Delta S = S^L - S^S$

At the equilibrium melting temperature T_m, $\Delta G_v = 0$, so

$$\Delta S = \Delta H / T_m = L / T_m \qquad (1.2)$$

where L is the latent heat of fusion. Combining equations (1.1) and (1.2):

$$\Delta G_v = L - T(L/T_m), \text{ so for an undercooling}$$
$$\Delta T \text{ we can write}$$

$$\Delta G_v = L\Delta T / T_m \qquad (1.3)$$

This equation shows that the free energy decrease is greater the higher the degree of supercooling, and is a most useful result to which we will return.

Nucleation

Consider a given volume of liquid supercooled below T_m by a temperature interval ΔT. If a small sphere of solid forms (radius r), the free energy of the system will be lowered by an amount per unit volume corresponding to equation (1.3). Energy is required, however, to create the solid/liquid interface, of energy γ_{SL} per unit area. The free energy change in the system may be written:

$$\Delta G = -(4/3)\pi r^3 \Delta G_v + 4\pi r^2 \gamma_{SL} \qquad (1.4)$$

Figure 1.5 illustrates this relationship, where it may be seen that, for a given undercooling, there is a certain critical radius, r_c, of solid particle. Solid particles with $r < r_c$, known as embryos, will redissolve in the liquid to lower the free energy of the system, whereas particles with $r > r_c$, known as nuclei, will grow in order to decrease the energy of the system.

By differentiation of eqn 1.4 it can be shown that:

$$r_c = 2\gamma_{SL}/\Delta G_v \qquad (1.5)$$

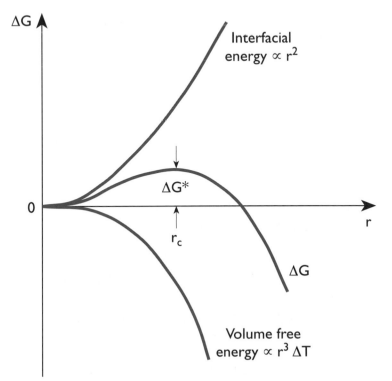

Fig. 1.5 The free energy change associated with the nucleation of a sphere of radius r.

and substituting eqn 1.3 gives:

$$r_c = (2\gamma_{SL} T_m / L)(1/\Delta T) \qquad (1.6)$$

indicating that r_c decreases with increasing undercooling.

For small degrees of undercooling, therefore, r_c is large and there is only a low probability that a large embryo will be formed in the liquid in a given time by random thermal motion of the atoms. There is thus likely to be only a low number of successful nuclei per unit volume of liquid. For high degrees of undercooling r_c is small, and the probability of forming such a nucleus is now very high, so that a high number of successful nuclei per unit volume of liquid will be observed. The implication of these effects upon the resultant microstructure will be considered next.

Growth of nuclei. Once stable nuclei are formed in the liquid, they grow at the expense of the surrounding liquid until the whole volume is solid. Most crystal nuclei are observed to grow more rapidly along certain crystallographic directions, causing spike-shaped crystals to develop. Further arms may branch out sideways from the primary spikes, resulting crystal with a three-dimensional array of branches known as *dendrites*, as shown in Fig. 1.6a.

Dendrites grow outwards from each crystal nucleus until they meet other dendrites from nearby nuclei. Growth then halts and the remaining liquid freezes in the gaps between the dendrite arms, as sketched in Fig. 1.6b. Each original nucleus thus produces a grain of its own, separated from the neighbouring grains by a *grain boundary*, which is a narrow transition region in which the atoms adjust themselves from the arrangement within one grain to that in the other orientation.

The *grain size* of a solidified liquid will thus depend on the number of nuclei formed, and thus on the degree of undercooling of the liquid. For example, if a liquid metal is poured into a (cold) mould, the layer of liquid next to the wall of the mould is cooled very rapidly. This gives rise to a very large local undercooling with the result that very many small nuclei of the solid are formed upon the mould wall, which grow to produce a very fine-grained layer of crystals (each perhaps less than 100 μm in size) at the surface of the casting, known as the 'chilled layer'.

The converse situation arises in nature over geological periods of time, when molten rock may cool very slowly and nucleation take place at small undercoolings. Few nuclei form, since r_c is so large, and beautiful mineral crystals of centimetre dimensions are commonly found. The grain size of a material is thus an important microstructural feature,

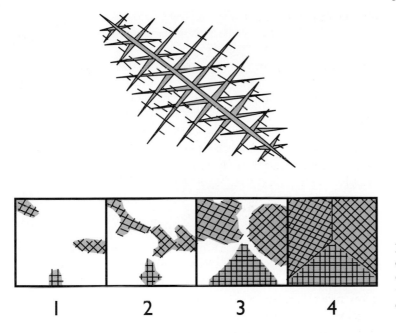

Fig. 1.6 (a) Drawing of a dendrite. (b) Schematic view of the freezing of a liquid by the nucleation and growth of dendrites.

and we will discuss later how its value may be controlled and what effect its magnitude may have upon the mechanical properties of the material.

An Introduction to Metallography

Let us next consider the various **techniques for microstructural examination.** It is usually necessary to prepare a section from a material in order to study the size, shape and distribution of crystals within it. In the case of metallic materials, this is referred to as *metallographic examination* ('materialography' is sometimes used more generally), but great precautions have to be taken at every stage to ensure that the method of preparation does not itself alter the microstructure originally present.

If the section for study is cut from the bulk by milling or sawing, or by the use of an abrasive cutting wheel, ample cooling and lubrication is provided to prevent its temperature from rising. Gross distortions from the cutting process are eliminated by grinding the surface with successively finer abrasives such as emery or silicon carbide. If the grains are coarse enough to be seen with the naked eye, one can at this stage prepare the surface for *macroscopic examination*.

This involves etching the surface of the specimen, usually in a dilute acid, by immersing it or swabbing it until the individual grains are revealed. Due to the different rates of

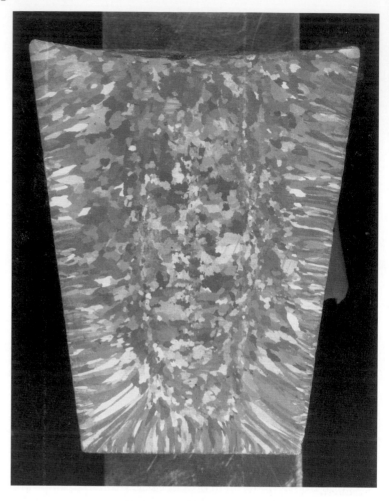

Fig. 1.7 The macrostructure of a cast metal. Large equiaxed grains have formed at the centre of the ingot. (Courtesy Dr P. A. Withey.)

chemical attack along different planes in a crystal, when the surface is etched, crystallographic terraces are formed upon each grain which reflect light in directions which vary with the orientation of the grain, so that some crystals appear light and some dark to the eye. The macrostructure of a piece of cast metal which has been prepared in this way is shown in Fig. 1.7.

Here, the crystals on the inside of the chilled layer have grown inwards to form long columnar crystals whose axis is parallel to the direction of heat flow. In contrast to this casting, in most engineering materials the grain size is too fine to be discerned without the use of a microscope, and specimen preparation is much more critical than for macro-examination. Polishing to a mirror finish is necessary, usually by holding the specimen against a horizontal rotating wheel covered with a short-pile cloth fed with a suspension or cream of a polishing agent. The latter can be magnesium oxide or aluminium oxide powder, although diamond pastes (of micrometre particle size) are commonly used. In the case of electrically conducting specimens such as metals, the final finish is often achieved by *electrolytic polishing*, where the specimen is made the anode in a suitable electrolyte. If the current density is correct, a bright scratch-free surface can be produced.

A much lighter etching treatment is applied for microscopical examination than for macro-studies. With some etching reagents and very short etching times, metal is dissolved only at the grain boundaries, giving rise to shallow grooves there, which are seen as a network of dark lines under the microscope.

A reflecting optical microscope may give magnifications of over $1000\times$, with a resolution of about 1 μm. The upper limit of magnification of the optical microscope is often inadequate to resolve structural features which are important in engineering materials, however, and *electron microscopy* is widely employed for this purpose. *Field-ion microscopy* is a research tool with a resolving power which permits the resolution of the individual atoms in crystals, which can be identified in such instruments by the *atom-probe* technique.

The two most commonly employed techniques of electron microcopy (EM) are SEM, *scanning electron microscopy* and TEM, *transmission electron microscopy*.

SEM

A schematic diagram of an SEM is given in Fig. 1.8. The beam produced by the electron gun is condensed and demagnified by the electromagnetic lenses to produce a 'probe' which is scanned over the surface of the specimen. Electrons emitted from the specimen surface are collected and amplified to form a video signal for a cathode-ray tube display. Typical resolutions of 10 nm may be obtained, with a depth of focus of several mm. It is this combination of high resolution with a large depth of focus that makes the SEM well suited to examine the fracture surfaces of samples.

TEM

A schematic diagram of a TEM is given in Fig. 1.9. Again the system is enclosed in a very high vacuum, and the image is viewed by focusing electrons upon a fluorescent screen after their transmission through the specimen. A resolving power down to 180 pm is obtainable in modern instruments. Two types of specimen may be studied: replicas and foils.

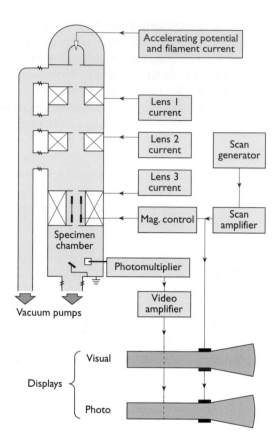

Fig. 1.8 Schematic diagram of a scanning electron microscope.

Replicas. After polishing a specimen as for optical microscopy, the surface is etched to reveal the required metallographic detail and produce surface relief. The surface is then overlaid with a cellulose acetate or similar film which, when stripped, replicates the surface relief (Fig. 1.10a). The stripped replica can be coated by evaporation with carbon and 'shadowed' with a heavy metal such as gold or platinum, which gives enhanced image contrast in the TEM after the acetate is removed with a solvating reagent.

In the case of specimens containing small particles of a different phase, such as precipitates of carbides in steel, it may be possible to retain in the replica, when it is stripped from the specimen, the actual particles which originally lay in the polished surface of the specimen (Fig. 1.10b). These are known as 'extraction replicas' and they permit the chemical and crystal structure of the precipitates to be analysed in the TEM (see below).

Foils. A much wider use is made in the TEM of very thin (∼ 500 nm) samples, which may be produced by a variety of methods. The most widely used is electrolytic thinning of the material, so that it cannot suffer mechanical damage. If this technique cannot be applied (as in the case of ceramic materials, for example) a useful alternative method of preparation is to bombard the surfaces of a slice of the material with energetic ions (≤ 10 keV), usually argon. The total time required to produce the final foil by this method is usually greater by at least a factor of five than the electrochemical method.

Material Analysis by Electron Microscopy

Electron microscopes are capable of providing three types of analysis, namely *visual* analysis of the microstructure at high resolution, *structural* analysis of the crystals themselves from electron diffraction measurements, and *chemical* analysis relying on efficient detection and discrimination of X-rays emitted from the specimen when bombarded with high energy electrons in the microscope.

In recent decades, high resolution microanalytical scanning transmission electron microscopes have been developed. In these instruments, electron beams with accelerating voltages between 100 and 400 keV can be focused down to provide chemical analysis on a scale of 10s of nm, providing information leading to an improved understanding of microstructures in a wide range of engineering materials.

A guide to further reading in this highly specialised area is given at the end of the Chapter.

Some Simple Phase Diagrams

The microstructures revealed by the above techniques are most easily understood by reference to the relevant *phase diagram*. In the case of a system with two components (*a binary system*) the phase diagram consists of a two-dimensional plot of temperature versus composition, which marks out the composition limits of the phases as functions of the temperature. We will now introduce some simple examples of phase diagrams, which we will correlate with some microstructures.

Solid Solubility

In a solid solution the crystal structure is the same as that of the parent element – the atoms of the solute element are

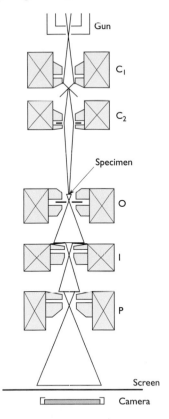

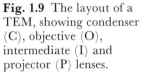

Fig. 1.9 The layout of a TEM, showing condenser (C), objective (O), intermediate (I) and projector (P) lenses.

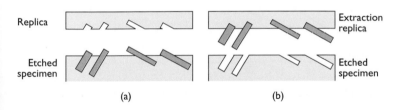

Fig. 1.10 Principle of (a) surface replication and (b) extraction replica.

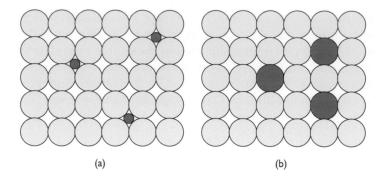

Fig. 1.11 Solid solutions formed (a) interstitially and (b) substitutionally.

(a)   (b)

distributed throughout each crystal, and a range of composition is possible. The solution may be formed in two ways: (a) in *interstitial* solid solutions the atoms of the solute element are small enough to fit into the spaces *between* the parent material atoms, as illustrated in Fig. 1.11(a). Because of the atom size limitation involved, interstitial solid solutions are not common, although, in steel, carbon atoms dissolve to a limited extent in iron crystals in this way.

(b) in *substitutional* solid solutions the atoms share a single common array of atomic sites (Fig. 1.11b). A copper crystal can dissolve up to approximately 35% zinc atoms in this way to form brass.

A few pairs of metals are completely miscible in the solid state and are said to form a 'continuous solid solution'; copper and nickel behave in this way and the phase diagram for this system is shown in Fig. 1.12. The horizontal scale shows the variation in composition in weight per cent nickel and the vertical scale is the temperature in °C. The diagram is divided into three 'phase fields' by two lines – the upper phase boundary line is known as the *liquidus* and the lower line as

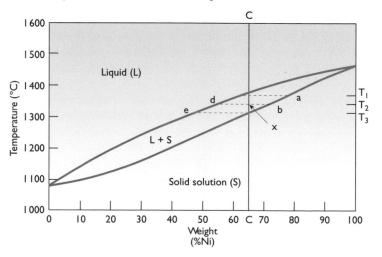

Fig. 1.12 The copper-nickel equilibrium diagram.

the *solidus*. At temperatures above the liquidus line, alloys of all compositions from pure copper to pure nickel will be liquid, while at temperatures below the solidus line, all alloys are in the solid state. It will be apparent that, unlike pure metals, alloys freeze over a *range* of temperature whose magnitude depends upon the composition of the alloy and is equal to the vertical separation of the liquidus and solidus at a given composition.

In working from a phase diagram, the beginner should always first consider some *specific composition* of alloy, and study its behaviour with respect to change in temperature. There is an important nickel–copper alloy known as Monel, whose retention of strength at high temperatures enables it to be used for turbine blading: its composition is approximately 65 weight per cent nickel – 35 weight per cent copper, and the vertical line *cc* in Fig. 1.12 has been constructed to correspond to this. Let us consider the solidification of a casting of this alloy, with molten metal being contained within a mould.

Considering a slow progressive decrease in temperature, at temperatures above T_1 the liquid phase is stable, but at T_1 solidification commences, and the two-phase field (marked $L + S$ in Fig. 1.12) of the diagram is entered. In any two-phase field of a phase diagram, the *compositions* of the two phases co-existing at a given temperature are obtained by drawing a horizontal (or *isothermal*) line. The required compositions are given by the intersections of the isotherm with the phase boundary lines. In the present case the isotherms are shown as dotted lines in Fig. 1.12, and at temperature T_1, liquid of composition *c* starts to freeze by depositing crystal nuclei of solid solution composition *a*, obtained by drawing the isothermal line at temperature T_1 in the two-phase field. As the temperature continues to fall, the loss of this nickel-rich solid causes the liquid's composition to become richer in copper, which it does by following the line of the liquidus, so that when temperature T_2 is reached, the composition of the liquid (given by the new isotherm) is now seen to be *d*. The growing crystals (normally in the form of dendrites, Fig. 1.6a) remain homogeneous, providing the temperature is not falling too quickly, and their composition follows the line of the solidus as they cool until, at temperature T_2, their composition is given by *b*. This crystal growth occurs by the deposition of layers of atoms which are richer in copper content, but atomic migration takes place by *solid state diffusion* within each dendrite between the new layers and the original nucleus, to enable the composition to adjust itself to *b*.

The dendrites we are considering will be at a temperature very close to their melting point, so that this diffusion process can continue to allow the dendrites to adjust their composition to follow the line of the solidus as the temperature continues to fall slowly – the remaining 'mother liquor' following the line of the liquidus. When temperature T_3 is reached, the last liquid (of composition e) freezes, and the accompanying solid-state diffusion brings the now completely frozen solid to the composition c once again. The solidified alloy is now (below T_3) is in a single-phase field once more, and is thus stable at all lower temperatures.

In summary, therefore, we see that in the slow solidification of a solid solution alloy, although we started with a liquid alloy of composition c and finished with a set of solid crystals of composition c, the process was more complicated than in the simple freezing of a pure solid. The initial nuclei were seen to have a different composition from the liquid in which they formed, and both the liquid phase and the solid phase progressively change their composition during the process of solidification.

The Lever Rule

In the temperature range T_1–T_3, when the two phases (L + S) were present, the construction of isothermal lines was shown (Fig. 1.12) to give the *composition* of the two phases which were in equilibrium. This same construction also determines how *much* of each phase is present at a given temperature, for a given alloy. Consider again the Monel of composition c; if at temperature T_2 the fraction of the alloy which is liquid is f_L, and the fraction of the alloy which is solid is f_S, then

$$f_L + f_S = 1.$$

If the concentration of nickel in the liquid phase $= d$ and the concentration of nickel in the solid phase $= b$, then

$$b.f_S + d.f_L = c,$$

but
$$f_S = 1 - f_L,$$

so
$$c = b - b.f_L + f_L + d.f_L, \text{ i.e.}$$

$$f_L = \frac{b - c}{b - d}$$

$$f_S = \frac{c - d}{b - d}$$

at this temperature. These relationships are known as the 'lever rule' because an isothermal 'tie-line' within a two-phase region may be considered as a lever of length bd whose fulcrum is at the point x (Fig. 1.12) where the line representing the composition (c) of the alloy intersects the isothermal line. The fraction of a phase having a composition indicated by *one* end of the lever is equal to the ratio of the length of the lever on the *far side* of the fulcrum to the total lever length.

This construction is applicable to *all* two-phase regions on phase diagrams – e.g. to the diagrams to be discussed below which contain regions where two *solid* phases co-exist. The lever rule is of great value to the metallographer is assessing the approximate composition of alloys from the relative proportion of the phases present that are observed in the microscope.

Non-Equilibrium Conditions

If the above cast sample of Monel were cooled at a fast rate in the mould, the solid-state diffusion processes described above may require too long a time to complete, so that the composition changes cannot conform to the solidus. Diffusion in a liquid can, however, take place more readily so that the composition of the liquid may be assumed still to conform to the liquidus. Let us consider in Fig. 1.13 the solidification process under these conditions: at temperature T_1 the liquid of composition c will first deposit crystals of composition a as before. As the temperature falls to T_2, the liquid composition will follow the liquidus to d, but the layer of solid crystal (composition b) deposited at this temperature will not have had time at this fast rate of cooling to inter-diffuse with the nickel-rich material beneath, so that the 'average' composition of the dendrite will be given by b', and a concentration gradient will exist in the crystal. Similarly at T_3 the liquid will be of composition e, the crystal *surface* will be of composition f, but the average crystal composition will be f' (due again to inadequate time for diffusion). Solidification will not be complete until T_4, when the last interdendritic liquid of composition g is frozen to solid h: this brings the *average* composition of the solid to c, the starting composition.

The locus of the solidus line is thus *depressed* (along a, b', f', etc) compared with its position under equilibrium conditions (along a, b, f etc), and secondly the structure of the resulting solid is now inhomogeneous and is said to be *cored*. Each crystal will consist of layers of changing composition – the 'arms' of the original dendrite being richer in the higher-melting constituent (in this case, nickel) than the average,

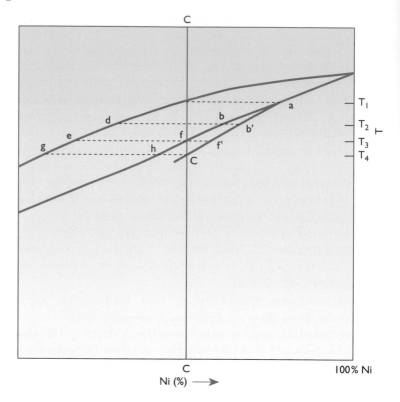

Fig. 1.13 Depression of the solidus by rapid solidification.

and the interdendritic regions being richer in the other constituent (i.e. copper) than the average. In a microsection of this structure, therefore, each grain will show a chemical heterogeneity which will be reflected in its rate of attack by the etchant, and Fig. 1.14 illustrates this effect in a sample of

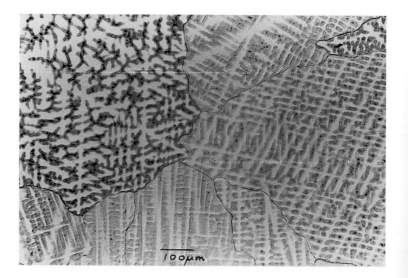

Fig. 1.14 The cored microstructure of a rapidly cooled solid solution of 30% zinc in copper. (Courtesy of the Copper Development Association.)

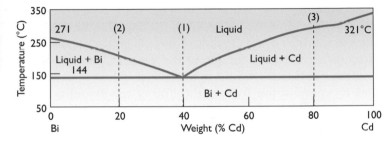

Fig. 1.15 The cadmium-bismuth equilibrium diagram.

chill-cast (i.e. rapidly solidified) brass which is a solid solution of 30 weight per cent zinc in copper. Depression of the solidus and 'coring' are common features of many cast alloys.

If the cored structure is undesirable, it may be removed by long heat treatments at high temperatures (known as 'homogenisation treatments') which allow the solute atoms to be redistributed by solid state diffusion.

No Mutual Solid Solubility (Simple Eutectic)

The cadmium-bismuth system is a simple eutectic system (see Fig. 1.15) which exhibits no solubility of cadmium in bismuth or of bismuth in cadmium. The phase diagram therefore consists of a liquidus line showing a minimum at the eutectic temperature, which is itself marked by a horizontal line. Since the solid phases formed consist simply of pure cadmium or pure bismuth, the *solidus* lines are coincident with the two vertical temperature axes.

Consider first the solidification of an alloy containing 40 wt% cadmium (alloy 1 in Fig. 1.15): it is liquid at temperatures above 144°C, and on cooling to this temperature it freezes isothermally to give an intimate mixture of cadmium and bismuth crystals known as a 'eutectic mixture' with the individual crystals in the form of plates or rods or small particles. Such a structure is sketched in Fig. 1.16a.

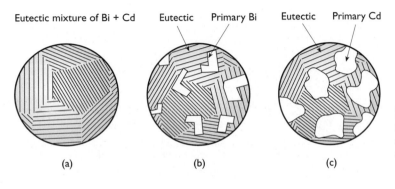

Fig. 1.16 Sketches of microstructures of Cd–Bi alloys of composition (a) 40% Cd, (b) 20% Cd and (c) 80% Cd.

Considering next alloy 2 in Fig. 1.15 which contains 20 wt% of cadmium, on crossing the liquidus line this will start to solidify, when crystals of pure bismuth will separate (the isothermal only intersects the vertical, pure bismuth, solidus), causing the liquid to become enriched in cadmium, and its composition follows the line of the liquidus as the temperature falls. At 144°C the bismuth crystals will be in equilibrium with liquid which has achieved eutectic composition: the liquid then freezes to form a eutectic mixture of crystals, giving the microstructure illustrated in Fig. 1.16b. Figure 1.16c illustrates the microstructure of alloy 3 in Fig. 1.15.

Limited Mutual Solid Solubility

A Eutectic System

Soft solders are based on lead and tin, and this system forms a eutectic system of this type, as shown in Fig. 1.17. Here the liquidus *ecf* shows a eutectic minimum at *c*, which means that an alloy containing 38 wt% lead will remain liquid to a relatively low temperature (183°C), and this is the basis of *tinman's solder*, which may be used for assembling electrical circuits with less risk of damaging delicate components through overheating them.

ea and *fb* are the solidus lines in Fig. 1.17; the lead-rich solid solution is labelled α phase and the tin-rich solid solution is termed the β phase. In interpreting the microstructures produced when alloys of various compositions are allowed to solidify, the reasoning will be a combination of those presented above.

Alloys with a tin content between 0 and *a* in Fig. 1.17 and between *b* and 100 will simply freeze to the single phase α and β solid solutions respectively when the temperature falls slowly. Following the previous reasoning, for a given alloy composition, solidification will start when the liquidus is

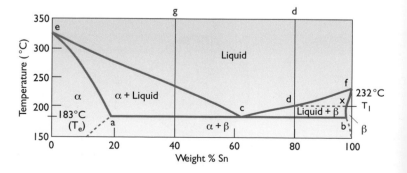

Fig. 1.17 The lead–tin equilibrium diagram.

crossed and be completed when the appropriate solidus is crossed. An alloy of composition *c* will solidify at the eutectic temperature (183°C) to fine a finely divided mixture of the α and β crystals.

However, considering the solidification of alloy *d* (Fig. 1.17), at temperature T_1, β crystals of composition *x* will nucleate and as the temperature falls towards the eutectic (T_e) the β crystals grow and change their composition along the solidus *fb* as the liquid phase composition follows the line *dc*. When the temperature reaches T_e, β crystals of composition *b* are in equilibrium with liquid of eutectic composition. This liquid then freezes to an α/β mixture and the microstructure will appear as in Figs 1.16b and 1.16c, except that the primary phase will consist of dendrites of a solid solution instead of a pure metal.

Non-Equilibrium Conditions

If the liquid alloy is allowed to cool too quickly for equilibrium to be maintained by diffusional processes, one might expect to observe cored dendrites of α or β phase, as discussed earlier. The depression of the solidus line under these conditions may, however, give rise to a further non-equilibrium microstructural effect if the composition of the alloy is approaching the limit of equilibrium solid solubility (e.g. *g* in Fig. 1.18).

If, due to rapid cooling, the solidus line is depressed from *ba* to *ba'*, alloy *g* would show some eutectic in its structure, whereas under conditions of slow cooling it would simply

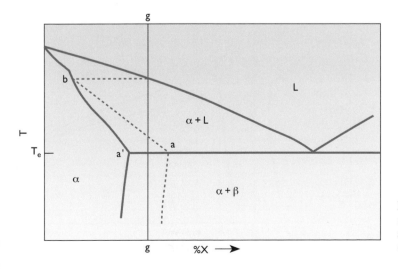

Fig. 1.18 Formation of metastable eutectic by rapid solidification of alloy of composition g–g.

freeze to a single phase, as predicted by the equilibrium phase diagram. An experienced metallographer can usually identify this effect, which is quite common in metal castings. In cast tin bronzes, for example, which are essentially copper-tin alloys, particles of hard second phases are often present (which can improve the mechanical properties of the material), even though the equilibrium phases diagram would predict a single-phase copper-rich solid solution for the compositions of the commonly used casting alloys.

A peritectic system

Figure 1.19 illustrates a second important way in which two solid solutions may be inter-related on a phase diagram. Temperature T_p is known as the *peritectic* temperature, and the boundaries of the β phase *fd* and *gd* are seen to come to a point at this temperature.

An alloy of composition *d* is said to have the peritectic composition, and we will now examine the nature of the phase change in more detail. Freezing of alloy *d* will start at temperature T_1 by the separation of crystals of the α solid solution; as the temperature falls under equilibrium conditions, the composition of the solid solution will follow the line of the solidus to *a* and that of the liquid will follow the line of the liquidus to *b*.

At T_p the liquid of composition *b* and the solid α phase of composition *a* react to form the solid solution β phase. Having for simplicity chosen an alloy composition corresponding to

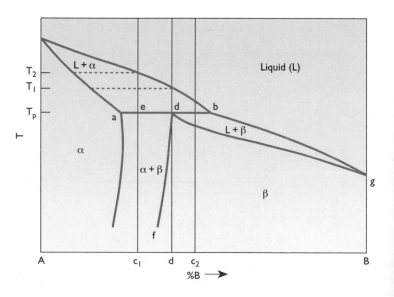

Fig. 1.19 A phase diagram showing a peritectic reaction.

the peritectic point, the reactants are fully consumed so that β is the only structure observed below the temperature T_p.

Considering now an alloy of composition c_1, this will likewise freeze initially at temperature T_2 to form α in the liquid phase, but at temperature T_p, although the reactants for the peritectic reaction are present (i.e. α phase of composition a and liquid of composition b), by the application of the lever rule it is seen that the fraction of solid phase present eb/ab is greater than that required for the peritectic reaction to proceed to completion, so that the β phase will be produced (with the disappearance of all the liquid phase and part of the α phase), and microstructure will consist of 'walls' of the peritectically produced β phase enveloping the unconsumed parts of the original α dendrites.

Non-equilibrium conditions
When a peritectic reaction begins, the α phase and the liquid are in contact and the β phase is formed at the solid-liquid interface, as illustrated in Fig. 1.20. When the β phase has formed an envelope about the α phase, the rate of reaction will depend upon the rate of diffusion of the reactants through the wall of β phase that separates them, and since this may be a sluggish process, it is quite commonly observed in cast alloys which have not been cooled extremely slowly, so that the peritectic reaction has not gone to completion and other '*metastable*' phases are seen.

Fig. 1.20 The microstructure of a specimen showing a peritectic reaction.

Phase Transformations in the Solid State

Changing Solid Solubility With Temperature

Figure 1.21 illustrates this feature in part of a eutectic diagram. Considering an alloy of composition c, this will solidify to a single-phase solid solution, α, which is stable only down to temperature T_1. The line ab is called a 'solid solubility' or 'solvus' line, and in the present example it shows that the solubility of X in the α phase falls from a value of $a\%$ at the eutectic temperature of $b\%$ at the lowest temperature on the ordinate axis.

As the temperature falls below T_1, the α phase crystals contain more of X than they would do at equilibrium – that is, they become *supersaturated*. If the cooling is slow, crystals of the β phase then form. Initial precipitation would take place along the grain boundaries of the original α phase – firstly because the atoms are more loosely held in the grain boundaries and so might be expected to 'break away' more readily to form the new phase, and secondly because the atomic disarray at the α-phase grain boundaries could help to accommodate any local volume changes associated with the growth of the new β crystals.

In many systems, the change of solubility with temperature is so great that the second phase cannot all be accommodated in the grain boundaries of the primary phase, and precipitation *within* the primary grains then occurs. This 'intragranular' precipitation is usually found to take the form of plates or needles in parallel array (see Fig. 1.22). This striking geometrical feature arises from the tendency of the new crystals to grow with their interfaces aligned parallel with certain definite crystal planes of the primary phase. These planes

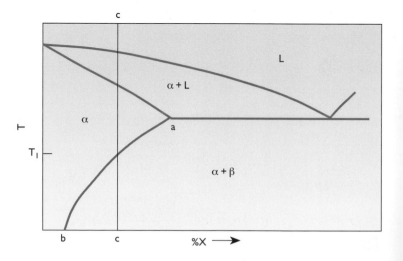

Fig. 1.21 Phase diagram showing decreasing solid solubility in the α-phase with decreasing temperature.

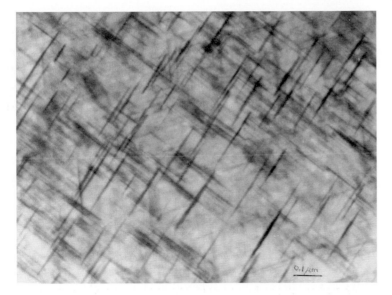
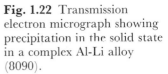

Fig. 1.22 Transmission electron micrograph showing precipitation in the solid state in a complex Al-Li alloy (8090).

will be such that there is a better atomic fit across the α/β interfaces than if the β phase was randomly distributed inside the α phase. Considering the critical nucleus size, if we apply equation 1.5 (which applies to a phase change from liquid to solid) to the present case of nucleation of one solid phase within another, these nuclei are in special orientations in order to minimise the value of $\gamma_{\alpha/\beta}$, the interfacial energy between the two phases.

Eutectoid and Peritectoid Processes

We have seen that eutectic and peritectic phase changes may proceed as follows:

eutectic: liquid phase $\underset{\text{heating}}{\overset{\text{cooling}}{\rightleftharpoons}}$ solid A + solid B

peritectic liquid phase + solid A $\underset{\text{heating}}{\overset{\text{cooling}}{\rightleftharpoons}}$ solid B

Processes analogous to these can take place wholly in the solid state, and are then called 'eutectoid' and 'peritectoid' respectively:

eutectoid: solid A $\underset{\text{heating}}{\overset{\text{cooling}}{\rightleftharpoons}}$ solid B + solid C

peritectoid solid A + solid B $\underset{\text{heating}}{\overset{\text{cooling}}{\rightleftharpoons}}$ solid C

No new principles are involved in these changes, and we will not discuss them in detail here, although it must be emphasised that, being wholly in the solid state, and dependent upon atomic migration to proceed, they may well be suppressed if the material is cooled quickly.

Quenched Structures

Alloys may deliberately have their microstructure controlled by cooling the material very rapidly ('quenching') to room temperature from a high temperature, single phase region of the phase diagram, such as the α field in Fig. 1.21. This may involve plunging the piece straight from a furnace into a bath of oil or water, and in general two possibilities exist:

1. A metastable *supersaturated α solid solution* may form. If the temperature is then progressively raised again until solid state diffusion may proceed at a measurable rate, the supersaturation will be relieved by the formation of a second phase. These are the processes underlying 'age harding', which we will discuss more fully later.

2. In some systems, the instability of the α phase is so high at low temperature, that it undergoes a *diffusionless* phase transformation to a **different** (again, metastable) structure we will call α'. This type of phase change is called a *martensitic* phase change, and its most important occurrence is in steel – to which we will return to consider in some detail.

Some alloys have been designed to transform to a martensitic phase when they are plastically deformed. If they are then heated, they may revert completely to their original crystal structure: this is accompanied by a reversal of the original plastic deformation, a process known as the **shape-memory effect**. A number of shape-memory alloys (SMA) exist: the earliest devices were of Ni–Ti ('Nitinol'), but more recently Cu– and Fe–based SMAs have been developed in which an element 'remembers' its shape prior to deformation. During this shape recovery, the element can produce a displacement as a function of temperature or, if constrained, a force *and* displacement. SMAs have been employed in circuit breaker actuators as well as in certain prosthetic devices for the fixation of artificial teeth.

THE MOLECULAR STRUCTURE OF ORGANIC POLYMERS AND GLASSES

The highly regular crystalline structures we have considered so far will not be formed if the interatomic binding requirements are satisfied simply by adding new units to the end of a chain. This can lead to the formation of high molecular weight, long chain structures known as **polymers.**

Although their properties differ widely, all polymers are made up of long molecules with a covalently bonded chain of atoms forming a 'backbone'. In organic polymers the chain is of cabon atoms, but in other polymers the chain could be of oxygen or silicon atoms, for example. The *configuration* of the polymer molecule is the arrangements of atoms which cannot be altered except by breaking primary chemical bonds; the *conformations* arc atomic arrangements which can be altered by rotating groups of atoms around single backbone bonds

The simplest structure of this kind to consider is the linear chain of polyethylene, which is the material used in plastic shopping bags, for example. The molecule of ethylene, C_2H_4, which is the *monomer* in this case, consists of the tetravalent carbon atoms forming strong covalent bonds with two hydrogen atoms, leaving a double bond between the carbon atoms, i.e. $CH_2 = CH_2$. Polymerisation breaks the double bond, allowing it to link with other activated ethylene monomers forming a long chain or *macromolecule*. The ends of the chain either link with other macromolecules or end with a *terminator*, such as an –OH group.

In order to form solids with useful mechanical properties the polymers chains must be long: such *high polymers* may contain between 10^3 to 10^5 monomer units in a molecule, this number being known as the *degree of polymerisation (DP)*. In all commercial polymers there is a range of DP and thus of molecular lengths, so one can only speak of an *average* molecular weight $\bar{M}$ in such materials. Such a value is usually described in terms of either the number or weight fraction of molecules of a given weight. The *number average* molecular weight $\bar{M}_n$ represents the total weight of material divided by the total number of molecules:

$$\bar{M}_n = \frac{\sum_{i=1}^{\infty} N_i M_i}{\sum_{i=1}^{\infty} N_i} = \sum_{i=1}^{\infty} X_i M_i$$

The *weight average* molecular weight reflects the weight of material of each size rather than their number:

$$\bar{M}_w = \frac{\sum_{i=1}^{\infty} N_i M_i^2}{\sum_{i=1}^{\infty} N_i M_i} = \sum_{i=1}^{\infty} w_i M_i$$

Figure 1.23 shows an example of the molecular weight distribution for all the chains in a polymer, showing the location of both $\bar{M}_n$ and $\bar{M}_w$. The usual technique employed to find this distribution is known as **Gel permeation chromatography (GPC)**.

The use of a GPC machine firstly involves dissolving the polymer in a suitable solvent. The solution, at a chosen concentration and temperature, is then introduced into the top of a column containing a gel formed of tiny spherical porous particles. The solution seeps down the column, and as this happens the individual polymer molecules diffuse into the pores of the gel. The longer molecules cannot readily do this, but the shorter molecules can penetrate the pores. Thus the larger the molecule, the less time it spends inside the gel, and the sooner it flows through the column. On their way down the column, the larger molecules get ahead of the smaller ones, and by the time the whole column is traversed the molecular masses are sufficiently separated for their distribution to be measured as illustrated in Figure 1.23. Successive samples of solution are collected, and the weight fraction of polymer in each sample is estimated from measurements of its refractive index. The apparatus has to be calibrated

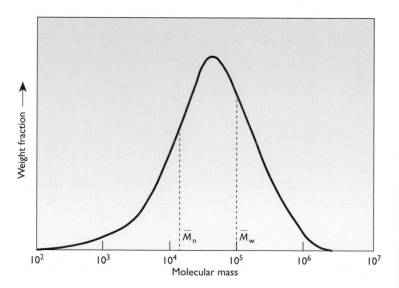

Fig. 1.23 Molecular mass distribution for high density polyethylene.

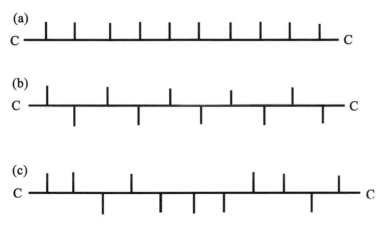

(a)

(b)

(c)

Fig. 1.24 Schematic representation of the arrangement of side groups in linear polymers with carbon chain C–C: (a) isotactic, (b) syndiotactic and (c) atactic.

by measuring the time for a solution of known moleclar weight to traverse the column.

The simple linear chain of polyethylene may have its chemical constitution modified in order to produce materials with different properties. By replacing one or two H atoms of the monomer by a side-group or radical, the *vinyl* group of polymers is formed: $-(CH_2-CXY)_n-$. If **X** is H and **Y** is Cl, polyvinyl chloride is produced, CH_3 substitution for **Y** produces polypropylene and C_6H_5 gives polystyrene. If **X** is CH_3 and **Y** is COO CH_3, polymethyl methacrylate (PMMA) is produced. These substitutions makes the monomer molecule asymmetrical so the polymer chain now can be formed in several ways:

(i) An *isotactic* linear polymer has all the side group on the same side of the chain (Fig. 1.24a).
(ii) A *syndiotactic* linear polymer has the side group alternating regularly on either side of the chain (Fig. 1.24b).
(iii) If the side groups alternate randomly, it is termed an *atactic* polymer (Fig. 1.24c).

Many apparently linear polymers are in fact *branched*, Fig. 1.25, as a result of subsidiary reactions occurring during polymerisation. Polyethylene is available with a wide range of structures and hence properties. The low density (LDPE) types are extensively branched, whilst high density polyethylene (HDPE) is essentially linear. Medium density (MDPE) types fall between these extremes, and grades are used according to their application, ranging from packaging for the flexible lower density types to semi-structural for the stiffer high density polyethylenes.

Although the directions of the C–C bonds in the chain are

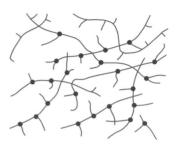

Fig. 1.25 Diagram of a branched linear polymer.

rigidly fixed, rotation about these bonds is relatively easy. The chain is thus flexible, and the bulk plastic consists of a tangled network of highly kinked chains which are locally linked together by the much weaker secondary bonds of the van der Waals or hydrogen bond type. When the temperature is raised these secondary bonds melt so that the polymer can flow like a viscous liquid, allowing it to be formed. The material regains its strength, reversibly, when it is cooled, and such polymers are known as **thermoplastics.**

Thermosets are made by mixing two components (a *resin* and a *hardener*) which react and harden. During polymerisation, chemical bonds are formed which **cross-link** the polymer chains. These strong covalent bonds between adjacent chains produce polymers of higher rigidity than the thermoplastics, and they cannot be resoftened by heating once the network of cross-linking primary bonds has been established. A familiar example of this type of polymer is epoxy, which is used as an adhesive and as the matrix of fibre-glass composites.

Elastomers may be classified as linear thermoset polymers with occasional cross-links which enable the material to return to its original shape on loading. The common elastomers are based on the structure:

$$
\left(
\begin{array}{cccc}
H & & H & \\
| & & | & \\
-\;C\;-\;C\;=\;C\;-\;C\;- \\
| & | & | & | \\
H & H & R & H
\end{array}
\right)_n
$$

with the site 'R' occupied by CH_3 (in natural rubber), H (in synthetic rubber, or Cl (in neoprene, which is used for seals because of its oil-resistance).

Polymer Crystals

It is possible for some long-chain molecular solids to crystallise if the chains happen to be packed closely together. Molecules with ordered, regular structures with no bulky side groups and a minimum of chain branching will usually crystallise. Thus the chains of an isotactic polymer (with its side-groups, if any, symmetrically placed on the backbone) may, if slowly cooled, be pulled by the secondary bonds into parallel bundles, often forming **chain-folded** crystal structures (Fig. 1.26).

Even if this crystallinity is good enough to enable such a

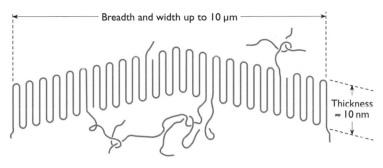

Fig. 1.26 Chain-folded lamellar crystal in polyethylene.

polymer to diffract X-rays, even the most crystalline of polymers is only 98% crystal. The crystalline parts are separated by amorphous regions. In melt-crystallised polymers, the chain-folded regions may grow 3-dimensionally into **spherulites**. These are highly crystalline units that grow with spherical symmetry to diameters of the order 0.01 mm until they impinge on one another. Spherulites scatter light easily, so that transparent polymers become translucent when crystalline, and under polarized light the spherulitic structure may be revealed (Fig. 1.27).

The degree of crystallinity may be determined from measurements of specific volume or density. Control of the degree of crystallinity must be exercised in the production of artefacts from these materials, because this exerts a strong influence on their mechanical properties. The highly crystalline form of a polymer will normally be stronger, though more

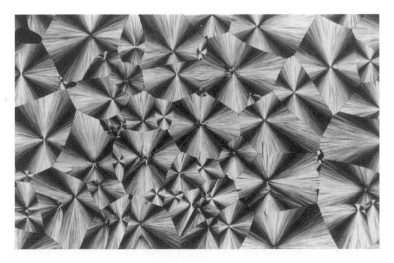

Fig. 1.27 A polarised light micrograph of spherulites of polyethylene oxide. The long edge of the micrograph corresponds to 1.4 mm on the sample (Courtesy of Professor C. Viney).

brittle, than one with low crystallinity. Spherulite size can be reduced by using nucleating agents, and tougher polymers are produced as a result.

We may therefore summarise the classification of organic polymers as follows:

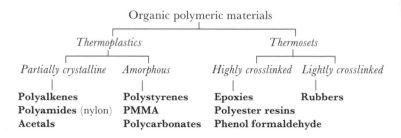

Organic polymeric materials

Thermoplastics		*Thermosets*	
Partially crystalline	*Amorphous*	*Highly crosslinked*	*Lightly crosslinked*
Polyalkenes	**Polystyrenes**	**Epoxies**	**Rubbers**
Polyamides (nylon)	**PMMA**	**Polyester resins**	
Acetals	**Polycarbonates**	**Phenol formaldehyde**	

Inorganic Glasses

Glasses are the most familiar inorganic polymers: the main difference between them and the long-chain organic materials described above is that the molecular chains in glass consist of more complex units based on the SiO_4 tetrahedral unit. This has a structure similar in shape to the diamond tetrahedron (Fig. 1.2a) which is highly stable and which can link with other tetrahedra by the sharing of an oxygen atom. Unlike organic polymers, the molecules are not constrained to form linear chains, but form three-dimensional random networks (Fig. 1.28). The network has a high degree of mobility at high temperature, to form a liquid which, in the case of pure silica, has a high viscosity.

Fig. 1.28 Silica tetrahedra connected in a random network to form a glass.

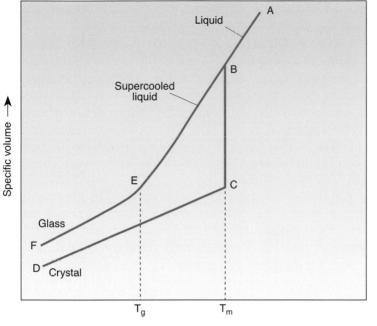

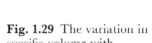

Fig. 1.29 The variation in specific volume with temperature for crystalline materials and glasses.

This problem is overcome in commercial glasses by introducing other metal oxides usually Na_2O and CaO which have effect of breaking up the network. These network modifiers reduce the cross-linking between the tetrahedra, making the glass much more easily worked at high temperature.

The way the volume of a given mass of this material changes as it is cooled is shown in Fig. 1.29. At A the material is a normal liquid: If it crystallises on cooling, then *B* represents the freezing point (T_m) at which there is a sharp decrease in the volume to *C*, after which the crystalline material will continue to shrink as the temperature falls, but at a slower rate, to *D*. In the case of a glass, which does not crystallise as it cools, shrinkage will occur along *AE* (Fig. 1.29). At a particular temperature, depending upon the rate of cooling, the rate of contraction slows to that along *EF*, whose slope is comparable to that of *CD* found in the crystalline material. The temperature at which the rate of contraction changes is known as the *glass transition temperature* (T_g), and its value depends on the rate of cooling of the glass – being lower at slower cooling rates.

The Glass Transition Temperature in Organic Polymers

In a polymer which does not crystallise, at low temperatures secondary bonds bind the molecules of a polymer into an amorphous solid, or glass. By analogy with the behaviour of inorganic glasses described above, above a certain temperature, known as the glass transition temperature (T_g), thermal energy causes the polymer molecules to rearrange continuously which in turn causes the volume of the polymer to increase. As the temperature rises a polymer becomes first *leathery* then *rubbery*, and eventually has the characteristrics of a viscous *liquid*. A curve of specific volume with temperature again appears as in Fig. 1.29, with an inflection appearing at T_g. The actual value of T_g again depends on the rate of temperature change, for example the lower the cooling rate the lower the value of T_g.

On the level of the molecular structure, the glass transition temperature is the temperature for a particular polymer at which molecular rotation about single bonds in the backbone becomes possible. Rotation is thermally activated, and the easier it is the lower the value of T_g for a particular polymer. T_g is thus increased with increasing strength of secondary bonds between chains, by cross-linking between chains, and by the presence of side-branches. *Plasticisers* reduce T_g, as they increase the space between chains, increasing chain mobility.

In the following Chapter we will review the ways in which the mechanical properties of engineering materials may be assessed.

FURTHER READING

D.A. Porter and K.E. Easterling, *Phase transformation in Metals and Alloys*, Chapman & Hall, 1992.

W.D. Kingery, H.F. Bowen and D.R. Uhlman, *Introduction to Ceramics*, Wiley, 1976.

N. G. McCrum, C.P. Buckley and C.B. Bucknall, *Principles of Polymer Engineering*, O.U.P., 1988.

P.E.J. Flewitt and R.K. Wild, *Microstructural Characterisation of Metals and Alloys*, The Institute of Materials, London, 1985.

Fred W. Billmeyer Jr., *Textbook of Polymer Science*, 3rd edition, John Wiley, NY, 1984.

2 The Determination of Mechanical Properties

Mechanical testing of engineering materials may be carried out for a number of reasons: The tests may simulate the service conditions of a material, so that the test results may be used to predict its *service performance*. Mechanical testing may also be conducted in order to provide engineering *design data*, as well *acceptability* – the main purpose of which is to check whether the material meets the specification.

In the USA, the American Society for Testing Materials (ASTM) publish standard specifications and methods of testing which are updated every three years. In the UK, the British Standards Institution (BSI) publish an annual catalogue of all BSI Standards, and which also refers to the agreed European Standards (EN series). All of these organisations issue publications relating to the selection of test-pieces and the conducting of mechanical tests. We will consider a number of these tests in turn.

THE TENSILE TEST

This is a widely used test for measuring the stiffness, strength and ductility of a material. The testing machine subjects the test-piece to an axial *elongation*, and the resultant *load* on the specimen is measured. Depending on the nature of the product being tested, the specimen may be round or rectangular in cross-section, and region between the grips is usually reduced in cross-section, on which the *gauge length* is marked, and upon which the extension is measured.

We will consider the response of a ductile metal as an illustration. The load – elongation data are normally converted to stress and strain:

Stress = Load/Cross-sectional area

Strain = Extension of gauge length/Original gauge length

and Fig. 2.1 illustrates the behaviour at small strains. The linear part of the curve may correspond to easily measured elongations in some polymeric materials, but in metals the displacements are very small and usually require the use of an

37

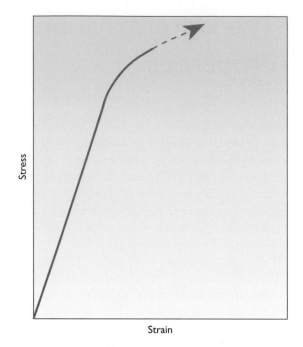

Fig. 2.1 Tensile test at small strains.

extensometer or *resistance strain-gauge* to measure them with sufficient accuracy.

This part of the curve is described by Hooke's Law, and it represents *elastic* behaviour. Its slope corresponds to Young's Modulus (E), which is given by the ratio of stress to strain. We have seen in Fig.1.1, which plots Young's Modulus vs. density for engineering materials, that the value of Young's Modulus can vary by over three orders of magnitude for different materials with elastomers having values of the order 0.1 GPa and metals and ceramics having values of hundreds of GPa.

It is clearly important for design engineers to know the stress at which elastic behaviour ceases. The *limit of proportionality* is the highest stress that can be applied with Hooke's Law being obeyed, and the *elastic limit* is the maximum stress which can be applied without causing permanent extension to the specimen. Neither of these stresses will be found in reference books of properties of materials, however, since their experimental measurement is fraught with difficulty. The more sensitive the strain gauge employed in the experiment, the lower the limit of proportionality and the elastic limit will appear to be. So as one changes from mechanical measurement of the strain with, say, a micrometer, to an optical lever device, and then to an electrical resistance strain gauge and finally to optical interferometry one would detect departure from elastic behaviour (as defined above) at progressively lower stresses, due to the presence of *microstrains*.

Departure from elasticity is therefore defined in an empirical way by means of a *proof stress*, whose value is independent of the accuracy of the strain-measuring device. Having constructed a stress-strain curve as in Fig. 2.1, an arbitrary small strain is chosen, say 0.1% or 0.2%, and a line parallel to Young's Modulus is constructed at this strain. The point of intersection of this line with the stress–strain curve defines the 0.1% or 0.2% *Proof Stress*, and values of this stress for different materials are available in books of reference since they provide an empirical measure of the limit of elastic behaviour. In the USA this stress is known as the *offset yield strength*.

Behaviour of Metals at Larger Strains

Figure 2.2 illustrates the form of a typical load-elongation curve for a ductile metal: after the initial elastic region, the gauge length of the specimen becomes plastic, so that if the load is reduced to zero the specimen will remain permanently deformed. The load required to produce continued plastic deformation increases with increasing elongation, i.e. the material *work hardens*.

The volume of the specimen remains constant during plastic deformation, so as the gauge length elongates its cross-sectional area is progressively reduced. At first, work hardening more than compensates for this reduction in area and the gauge length elongates uniformly. The *rate* of work

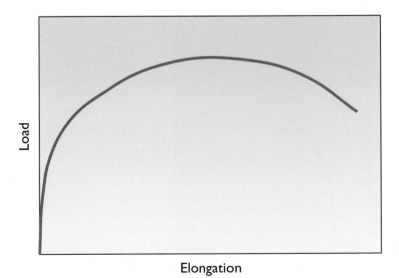

Fig. 2.2 Tensile test of a ductile metal.

hardening decreases with strain, however, and eventually a point is reached when there is an insufficient increase in load due to work hardening to compensate for the reduction in cross-section, so that all further plastic deformation will be concentrated in this region and the specimen will undergo *necking*, with a progressive fall in the load. The onset of necking is known as *plastic instability*, and during the remainder of the test the deformation becomes localised until fracture occurs.

The Engineering Stress–Strain Curve

The load–elongation curve of Fig. 2.2 may be converted into the Engineering Stress–Strain curve as shown in Fig. 2.3. The engineering, or conventional, stress σ is given by dividing the load (L) by the *original* cross-sectional area of the gauge length (A_o), and the strain (e) is as defined above, namely the extension of the gauge length $(l–l_o)$ divided by gauge length (l_o).

The maximum conventional stress in Fig. 2.3 known as the *Ultimate Tensile Stress (UTS)* defined as:

$$UTS = L_{max}/A_o$$

and this property is widely quoted to identify the strength of materials.

The tensile test also provides a measure of ductility. If the fractured test-piece is reassembled, the final length (l_f) and final cross-section (A_f) of the gauge length may be measured,

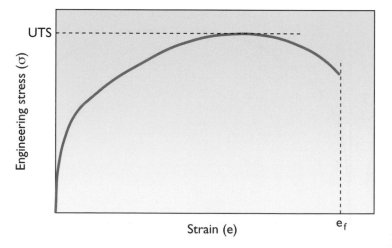

Fig. 2.3 Engineering stress–strain curve.

and ductility expressed either as the engineering strain at fracture:

$$e_f = (l_f - l_o)/l_o,$$

or the reduction is cross-section at fracture, RA, where:

$$RA = (A_o - A_f)/A_o$$

These quantities are usually expressed as percentages. Because much of the plastic deformation will be concentrated in the necked region of the gauge length, the value of e_f will depend on the magnitude of the gauge length – the smaller the gauge length the greater the contribution to e_f from the neck itself. The value of the gauge length should therefore be stated when recording the value of e_f.

The True Stress–Strain Curve

The fall in the engineering stress after the UTS is achieved, due to the presence of the neck, does not reflect the change in strength of the metal itself, which continues to work harden to fracture. If the *true stress*, based on the *actual* cross section (A) of the gauge length, is used, the stress–strain curve increases continuously to fracture, as indicated in Fig. 2.4.

The presence of a neck in the gauge length at large strains introduces local triaxial stresses that make it difficult to determine the true longitudinal tensile stress in this part of the curve. Correction factors have to be applied to eliminate this effect.

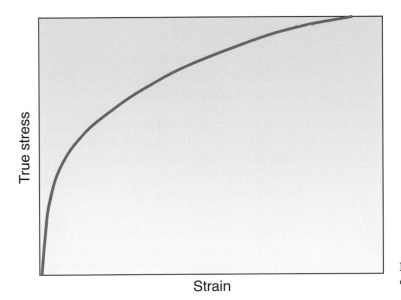

Fig. 2.4 True stress–strain curve.

Figure 2.4 illustrates the continued work hardening of the material until fracture, but, unlike the Engineering Stress-strain curve, the strain at the point of plastic instability, and the UTS are not apparent. Their values may be readily obtained from Fig. 2.4 by the following approach:

At the UTS, the load (L) passes through a maximum, i.e. $dL = 0$.

Since $L = \sigma A$, we may write

$$dL = \sigma \, dA + A \, d\sigma = 0$$

i.e.
$$- dA/A = d\sigma/\sigma \tag{2.1}$$

Since the volume of the gauge length $(V = A \, l)$ is constant throughout the test, $dV = 0$, and

$$dl/l = -dA/A \tag{2.2}$$

So from (2.1) and (2.2) we may write:

$$d\sigma/\sigma = dl/l \tag{2.3}$$

But the strain, e, is defined as

$$e = (l - l_o)/l_o = l/l_o - 1 \tag{2.4}$$

Therefore $de = dl/l_o = (dl/l)(l/l_o) \tag{2.5}$

So from (2.4) and (2.5) we obtain:

$$de = (dl/l)(1 + e) \tag{2.6}$$

Eliminating dl/l from (2.6) and (2.3) we find that, at the UTS:

$$d\sigma/de = \sigma_u/1 + e_u \tag{2.7}$$

where σ_u is the *true stress* at the UTS and e_u the strain at the point of plastic instability. Equation (2.7) thus enables us to identify these values by means of the *Considère construction*, whereby a tangent to the true stress-strain curve is drawn from a point corresponding to -1 on the strain axis (Fig. 2.5). Additionally, the intercept of this tangent on the stress axis will give the value of the UTS, since

σ_u /UTS $= A_o/A = l/l_o$ (since the volume is constant)

But from (2.4), $l/l_o = 1 + e$

so by similar triangles, it may be seen in Fig. 2.5 that the Considère tangent to the true stress-strain curve identifies both the point of plastic instability, the true stress at the UTS and the value of the UTS itself.

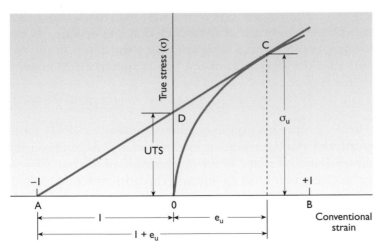

Fig. 2.5 The Considère construction.

BEND TESTING

Testing brittle materials in tension is difficult, due to the problems of gripping the specimens in a tensile machine without breaking them. Furthermore, in the case of materials such as glasses and ceramics, there is difficulty in shaping test-pieces of the 'dog-bone' shape without generating further flaws and defects in their surfaces.

Brittle materials are therefore frequently tested in bending in the form of parallel-sided bars, which are simple to make, and deformed in either three-point or four-point bending (Fig. 2.6). In three-point bending the maximum tensile stress occurs at a point opposite the central load and in four-point bending the whole of the surface between the central loading

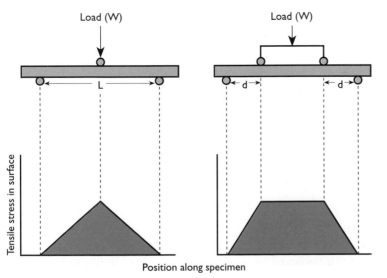

Fig. 2.6 The geometry of (a) three- and (b) four-point bend tests showing the corresponding stress distributions in the specimen surface.

edges, on the convex side of the bar, experiences the same maximum tensile stress. Provided that the spacing of the loading points is large compared with the depth of the bar and that the deflection of the bar is small, the maximum tensile surface stress is given by:

$$\sigma_{\max} = M/D \qquad (2.8)$$

where M is the maximum bending moment ($= WL/4$ for 3-point and $Wd/2$ for 4-point loading) and D depends on the dimensions and shape of the cross section of the bar. $D = \frac{1}{4} \pi r^3$ for a circular cross section and (breadth $\times$ depth^2)/6 for a rectangular section.

Because of the gradient of stress and strain through the cross-section of the specimen, from compressive on one side to tensile on the other, the apparent tensile strength values from a bend test tend to be higher than those from a tensile test. In brittle solids, the maximum tensile surface stress achieved in the test given by equation (2.8), is referred to as the **flexural strength,** or the **modulus of rupture**.

THE STATISTICS OF BRITTLE FRACTURE

Since brittle solids fail by the propagation of pre-existing cracks under tension, measurement of their tensile strength will be subject to considerable experimental scatter. Ceramics and glasses contain a distribution of crack lengths, so it follows that there will be a statistical variation in their measured strengths. Furthermore, a large component will fail at a lower stress than a small one, since it is more likely that it will contain one of the larger cracks – so there is a volume dependence of the strength. Statistical models of strength are therefore needed in order to give a true prediction of the mechanical properties of brittle solids.

Weibull has defined the *survival probability*, $P_s(V)$, as the fraction of identical samples (each of volume V) which survive loading to a tensile stress σ, proposing the relation:

$$P_s(V) = \exp\left\{-\frac{V}{V_o}\left(\frac{\sigma}{\sigma_o}\right)^m\right\} \qquad (2.9)$$

where σ_o, V_o and m are constants, m being known as the *Weibull modulus*. The lower the value of m, the greater the variability of strength. The values of these constants are found by experiment: if the stress varies with position (as in the case of a bend test), then equation 2.9 can be integrated over the volume to give the appropriate design load.

HARDNESS TESTING

Hardness is not a well-defined property of materials, and the tests employed assess differing combinations of the elastic, yielding and work-hardening characteristics. All the tests are essentially simple and rapid to carry out, and are virtually non-destructive, so they are well-suited as a means of quality control. The hardness of materials has been assessed by a wide variety of tests, but we will confine ourselves to discussing two types of measurement – the resistance to *indentation* and the *height of rebound* of a ball or hammer dropped from a given distance.

Indentation Hardness Tests

There are two types of indentation hardness tests. The first (Brinell and Vickers Hardness) measures the size of the impression left by an indenter of prescribed geometry under a known load, the second type (Rockwell) measures the depth of penetration of an indenter under specified conditions.

The Brinell Test

The surface of the material is indented by a hardened steel ball (whose diameter D is usually 10 mm) under a known load (L) (e.g. 3000 kg for steel), and the average diameter of the impression measured with a low-power microscope. The Brinell number (H_B) is the ratio of the load to the contact surface area of the indentation. Most machines have a set of tables for each loading force, from which the hardness may be read in units of kgf mm^{-2}.

If other sizes of indenter are used, the load is varied according to the relation: $L/D^2 = $ constant in order to obtain consistent results. The constant is 30 for steel, 10 for copper and 5 for aluminium.

The Vickers Test

This machine uses a diamond square-based pyramid of 136° angle as the indenter, which gives geometrically similar impressions under differing loads (which may range from 120 kg to 5 kg). A square indent is thus produced, and the user measures the average diagonal length and again reads the hardness number (H_V) from the tables.

The Brinell and Vickers hardness values are identical up to

a hardness of about 300 kgf mm^{-2}, but distortion of the steel ball occurs in Brinell tests on hard materials, so that the test is not reliable above values of 600 kgf mm^{-2}.

For steels there is a useful empirical relationship between the UTS (in MPa) and H_V (in kgf mm^{-2}.), namely:

$$\text{UTS} \approx 3.2 \; H_V$$

The Rockwell Test

This uses either a steel ball (Scale B) or a diamond cone (Scale C), and the indenter is first loaded with a minor load of 10 kg m, and the indicator for measuring the depth of the impression is set to zero. The appropriate major load is then applied, and after its removal the dial gauge records the depth of the impression in terms of Rockwell numbers.

The correlation between the Rockwell, Vickers and Brinell Hardnesses is given in standard reference books, and is indicated in Fig. 2.7.

It should be pointed out that, in the case of materials which exhibit time-dependence of elastic modulus or yield stress (for example, most polymers), the size of the indentation will increase with time, so their hardness value will depend on the duration of application of the load.

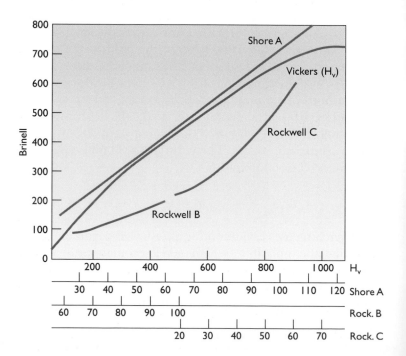

Fig. 2.7 The relation between various scales of hardness.

The Shore Scleroscope

This consists of a small diamond-pointed hammer weighing 2.5 g which falls freely from a standard height down a graduated glass tube. The height of the first rebound is taken as the index of hardness, and this simple apparatus may be readily transported for testing rolls, gears, etc. *in situ*. Its correlation with the indentation hardness techniques are also shown in Fig. 2.7.

The Shore test may also be employed to measure the elastic response of elastomers, as a check of the degree of crosslinking.

FRACTURE TOUGHNESS TESTING

The *toughness* of a material must be distinguished from its *ductility*. It is true that ductile materials are frequently tough, but toughness combines both strength and ductility, so that some soft metals like lead are too weak to be tough whereas glass-reinforced plastics are very tough although they exhibit little plastic strain.

One approach to toughness measurement is to measure the work done in breaking a specimen of the material, such as in the Charpy-type of impact test. Here a bar of material is broken by a swinging pendulum, and the energy lost by the pendulum in breaking the sample is obtained from the height of the swing after the sample is broken. A serious disadvantage of such tests is the difficulty of reproducibility of the experimental conditions by different investigators, so that impact tests can rarely be scaled up from laboratory to service conditions, and the data obtained can not be considered to be true material parameters.

Fracture toughness is now assessed by establishing the conditions under which a sharp crack will begin to propagate through the material, and a number of interrelated parameters may be employed express this property. To introduce these concepts we will first consider the *Griffith criterion* for the brittle fracture of a linear elastic solid in the form of an infinite plate of unit thickness subjected to a tensile stress σ, and having both ends clamped in fixed position. This is a condition of **plane stress**, where all stresses are acting in the plane of the plate. The elastic energy stored per unit volume is given by the area under the stress-strain curve, i.e. ½ stress × strain which may be written:

$$\text{Stored elastic energy} = \sigma^2/2E \text{ per unit volume}$$

where E is Young's modulus.

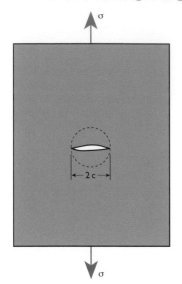

Fig. 2.8 Crack in plate of unit thickness.

If a crack is introduced perpendicular to σ, and the length of the crack $(2c)$ is small in comparison with the width of the plate, some relief of the elastic stress will take place. Taking the volume relieved of stress as a cylindrical volume of radius c (Fig. 2.8), then the elastic energy (U_v) released in creating the crack is:

$$U_v = (\pi c^2)(\sigma^2/2E)$$

in a uniform strain field, but is twice this if the true strain field is integrated to obtain a more accurate result, i.e.

$$U_v = \pi c^2 \sigma^2 / E \qquad (2.8)$$

Griffith's criterion states that, for a crack to grow, the release of elastic strain energy due to that growth has to be greater than the surface energy of the extra cracked surfaces thus formed. In Fig. 2.8, the area of crack faces is 4c, so if the surface energy of the material is γ per unit area, then the surface energy (U_s) required,

$$U_s = 4c\gamma \qquad (2.9)$$

Figure 2.9 represents energy versus crack length, and it is analogous to the diagram of free energy change versus embryo size in the theory of nucleation of phase transformations (Fig. 1.5). The critical crack size $c*$ is thus when:

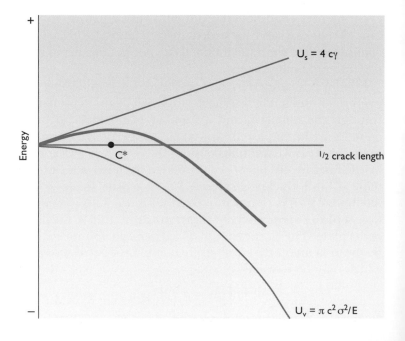

Fig. 2.9 Change in energy with crack length.

$$d/dc(4c\gamma - \pi c^2\sigma^2/E) = 0$$

i.e.

$$2\gamma - \pi c^*\sigma^2/E$$

so

$$\sigma = (2E\gamma/\pi c^*)^{\frac{1}{2}} \qquad (2.10)$$

which is the well-known Griffith equation . At stress σ, cracks of length below $2c^*$ will tend to close, and those greater than $2c^*$ will grow. Thus at a given stress level there is a critical flaw size which will lead to fracture, and conversely, for a crack of a given length, a critical threshold stress is required to propagate it.

Under conditions of **plane strain**, (2.10) becomes:

$$\sigma = [2E\gamma/\pi(1 - \nu^2)c^*]^{\frac{1}{2}} \qquad (2.10a)$$

where ν is Poisson's Ratio for the material.

In considering real materials, rather than an ideal elastic solid, the work required to create new crack surfaces is more than just the thermodynamic surface energy, 2γ. Other energy absorbing processes, such as plastic deformation, need to be included and these are taken into account by using the **toughness**, G_c, to replace 2γ in equations 2.10 and 2.10a, giving a fracture stress σ_F:

$$\sigma_F = (E G_c/\alpha\pi c)^{\frac{1}{2}} \qquad (2.11)$$

where $\alpha =$ unity in plane stress, and $(1 - \nu^2)$ in plane strain.

A related measure of toughness is the **fracture toughness**, K_c, which is related to G_c by:

$$G_c = \alpha K_c^2/E \qquad (2.12)$$

From equation (2.11) this may be written:

$$K_c = \sigma_F(\pi c)^{\frac{1}{2}} \qquad (2.13)$$

The fracture toughness is measured by loading a sample containing a deliberately introduced crack of length $2c$, recording the tensile stress σ_c at which the crack propagates. This is known *crack opening*, or **mode I** testing. K_c is then calculated from

$$K_c = Y\sigma_c(\pi c)^{\frac{1}{2}} \qquad (2.13a)$$

where Y is a geometric factor, near unity, which depends on the details of the sample geometry. The toughness can then be obtained by using equation (2.12).

This approach gives well-defined values for K_c and G_c for brittle materials (ceramics, glasses, and many polymers), but

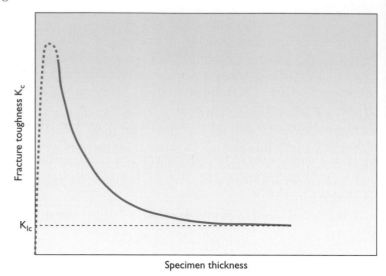

Fig 2.10 Change in fracture toughness with specimen thickness.

in ductile metals a *plastic zone* develops at the crack tip. The linear elastic stress analysis we have assumed so far can be applied only if the extent of plasticity is small compared with the specimen dimensions. Additionally, the testpiece must be sufficiently thick in order that most of the deformation occurs under conditions of *plane strain*. Figure 2.10 illustrates the effect of specimen thickness on the measured value of the fracture toughness, K_c: the fracture toughness can be halved as the stress conditions change from plane stress to plane strain with increasing specimen thickness. Once under plane strain, however, the value of the toughness becomes independent of the thickness, and this value is referred to as K_{Ic}, and is regarded as a material parameter. The corresponding value of G_{Ic} may be calculated from equation (2.12).

The two requirements of a relatively small plastic zone and plane strain conditions impose conditions upon the test-piece dimensions which have to be fulfilled in valid fracture toughness tests. These dimensions are always stated in the appropriate testing standards, e.g. BS 7448 (1991).

In circumstances of extensive plasticity, an alternative toughness parameter has been proposed, namely the **Crack Tip Opening Displacement (CTOD)**, usually given the symbol δ. The CTOD is obtained from the reading of a clip-gauge placed across the crack mouth, and under conditions of fracture a critical value of δ_c is determined. If the yield stress of the material is σ_y, the toughness may then be obtained from the relation:

$$G_c = \sigma_y \delta_c \qquad (2.14)$$

The J Integral

There are types of material, e.g. elastomers, which behave in a *non-linear* elastic manner, i.e. their reversible stress-strain graph is curved. The energy release rate for such materials is characterised by a parameter termed J, which is the non-linear equivalent of the potential energy release rate, G per unit thickness derived above. In a linear elastic material J would be identical to G, and the reader is directed to the British Standard BS 7448:1991, which describes methods for the determination of K_{Ic}, critical CTOD and critical J values of metallic materials.

TIME-DEPENDENT MECHANICAL PROPERTIES

We will now consider some material properties whose values are time-dependent, namely:

Creep, which refers to slow plastic deformation with time under load,
Fatigue, which is the damage and failure of materials under cyclic load,
Environment-Assisted Cracking in which cracks propagate under the *combined* action of an applied stress and an aggressive environment, and finally the **Time-dependent Elastic Properties**, which appears as a **viscoelastic behaviour** in polymers, and, under conditions of cyclic loading of solids in general, as the **damping capacity,** which measures the degree to which a material dissipates vibrational energy.

Creep

Creep occurs when materials are loaded above about $\frac{1}{3} T_m$. Tests are normally conducted under uniaxial tensile stress on a specimen similar to that used in tensile testing, and the test-piece and pull-rods may be situated in a tubular furnace whose temperature is accurately controlled. The strain in the specimen is monitored by a sensitive extensometer, and a typical tensile strain-time curves are shown in Fig. 2.11 for such experiments.

In curve A, after the initial instantaneous extension, a regime of decreasing creep-rate occurs (*primary creep*). *Secondary creep* occurs at an approximately constant rate, sometimes referred to as the *minimum creep rate*, which is followed by an accelerating regime of *tertiary creep*, leading to rupture.

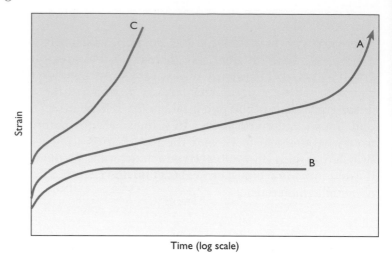

Fig. 2.11 Showing various strain–time curves for creep.

Curve *B* illustrates the type of result obtained if the test is conducted at a lower stress or lower temperature: only primary creep is observed, and fracture may not occur in the duration of the test. Curve *C* shows the effect of higher stresses or temperatures: secondary creep may be absent, and early failure may occur.

Returning to the general form of curve A in Fig. 2.11, the minimum creep rate ($\dot{\epsilon}$) under stress σ and temperature T may be characterised by the creep constants $\dot{\epsilon}_o$, σ_o, n and Q in the equation:

$$\dot{\epsilon} = \dot{\epsilon}_o (\sigma/\sigma_o)^n \exp -(Q/RT) \qquad (2.15)$$

Considerable time and thus expense is involved in determining creep data in this form, and design data are often given as a series of curves relating stress and time at a given test temperature to produce a given constant creep strain (1%, 2% etc.) or time to rupture (or 'creep life'), both are shown in the example of Fig. 2.12 for a nickel-based alloy in single crystal form (due to W. Schneider, J. Hammer and H. Mughrabi: *Superalloys 1992*, edited by S.D. Antolovich *et al.*, Warrendale, Pennsylvania , TMS, 1992, 589).

Acquisition of creep life data requires no strain measurement to be carried out, but it clearly can give no indication of the time spent in the various stages of the creep curve. Engineering components may thus be designed on a knowledge of the stresses which the relevant materials can withstand without fracture in times up to the anticipated service life. In the case of components for chemical and electricity generating plant, designs are generally based on the 100,000 hour rupture data, although economic benefits would

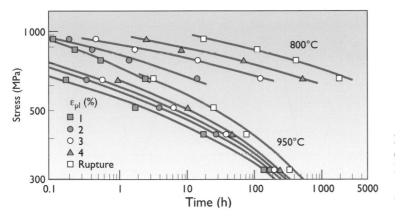

Fig. 2.12 Creep-rupture diagrams for a single crystal Ni-based superalloy (CMSX-4) for 800°C and 950°C.

obviously be derived if lives could be extended to, say, 250,000 hours.

For alloy development and production control, relatively short term creep tests are employed. Where a component experiences creep for very protracted periods, however, design data must itself be acquired from very lengthy tests rather than by extrapolation, since structural changes may occur in the material under these circumstances. Problems may also be encountered because the mechanisms of deformation and fracture may differ in different regimes of stress and temperature. With a limited number of materials, however, methods of extrapolating empirical data have been successfully developed for both creep strain and stress-rupture properties, allowing data for times of 10 years or more to be estimated from high-precision creep curves obtained in times of three months or less.

For further information on this topic, the reader is referred to R. W. Evans and B. Wilshire: *Creep of Metals and Alloys*, The Institute of Materials, London, 1985.

Fatigue

The progression of fatigue damage can be classified into a number of stages involving (a) the nucleation of microscopic cracks, (b) their growth and coalescence and (c) the propagation of a macroscopic crack until failure. The original approach to fatigue design involved characterising the *total fatigue life* to failure of initially uncracked testpieces, in terms of the number of applications of a cyclic stress range (the S–N curve in 'high cycle fatigue') or a cyclic strain range ('low cycle fatigue').

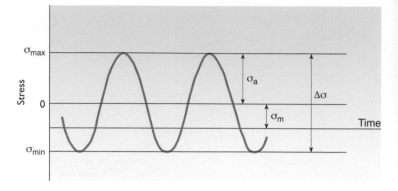

Fig. 2.13 A typical stress pattern in fatigue.

The S–N Curve

Smooth, un-notched testpieces are prepared, carefully avoiding sharp changes in cross-section which may give rise to stress concentrations. Testing machines apply fluctuating loads in tension, compression, torsion, bending, or combinations of such loads, and the number of cycles to specimen failure is recorded. Test methods are described in detail in the ASTM Standards E466–E468.

Figure 2.13 illustrates a typical stress pattern, the *stress range* ($\Delta\sigma = \sigma_{max} - \sigma_{min}$) and the *mean stress* ($\sigma_m = ($ [$\sigma_{max} + \sigma_{min}$]$/2$) being indicated, as well as the *cyclic stress amplitude*, σ_a ($= \Delta\sigma/2$). The mean stress may be zero, tensile or compressive in value, and typical test *frequencies* employed depend on the machine design – whether it is actuated mechanically or servo-hydraulically for example – but are typically in the range 10–100 Hz.

The data from such tests are usually plotted upon logarithmic axes, as illustrated in the S–N curve of Fig. 2.14. Because of the protracted time involved, tests are seldom conducted for more than 5×10^7 or 10^8 cycles, and the curves may be of two types – those showing a continuous decline and those showing a horizontal region at lives greater than about 10^6 cycles. The horizontal line thus defines a stress range below which the fatigue life is infinite, and this is defined as the *endurance limit* or *fatigue limit*. The majority of materials do not exhibit an endurance limit, although the phenomenon is encountered in many steels and a few aluminium alloys.

The curve may be described by:

$$\sigma_a = \sigma_f'(2N_f)^b \tag{2.16}$$

where σ_f' is the fatigue strength coefficient (which is roughly equal to the fracture stress in tension) and b the fatigue

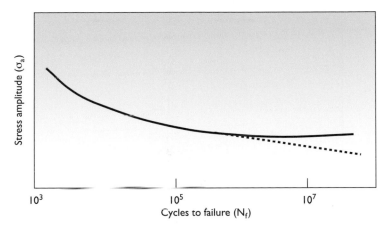

Stress amplitude (σ_a)

Cycles to failure (N_f)

10^3 10^5 10^7

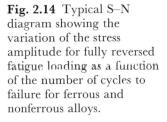

Fig. 2.14 Typical S–N diagram showing the variation of the stress amplitude for fully reversed fatigue loading as a function of the number of cycles to failure for ferrous and nonferrous alloys.

strength exponent, or *Basquin exponent*. For most metals $b \approx -0.1$, and some values of b and of σ_f' for a number of engineering alloys are given in Table 2.1.

Mean Stress Effects on Fatigue Life

The mean level of the imposed stress cycle (Fig. 2.13) influences the fatigue life of engineering materials: a decreasing fatigue life is observed with increasing mean stress value. The effect can be modelled by *constant life diagrams*, Fig. 2.15. In these models, different combinations of the mean stress and the stress range are plotted to provide a constant (chosen) fatigue life. For that life, the fatigue stress range for fully reversed loading ($\sigma_m = 0$) is plotted on the vertical axis at σ_{fo}, and the value of the yield strength σ_y and the UTS of the material is marked on the horizontal axis. The Goodman model predicts that, as the mean stress increases from zero, the fatigue stress for that life (σ_{fm}) decreases linearly to zero as

Table 2.1 Some cyclic strain-life data (C.C. Osgood: *Fatigue Design* New York, Pergamon Press, 1982)

Material	Condition	σ_y (MPa)	σ_f' (MPa)	ϵ_f'	b	c
Pure Al (1100)	annealed	97	193	1.80	−0.106	−0.69
Al-Cu (2014)	peak aged	462	848	0.42	−0.106	−0.65
Al-Mg (5456)	cold worked	234	724	0.46	−0.110	−0.67
Al-Zn-Mg (7075)	peak aged	469	1317	0.19	−0.126	−0.52
0.15%C steel (1015)	normalised	228	827	0.95	−0.110	−0.64
Ni-Cr-Mo steel (4340)	quenched and tempered	1172	1655	0.73	−0.076	−0.62

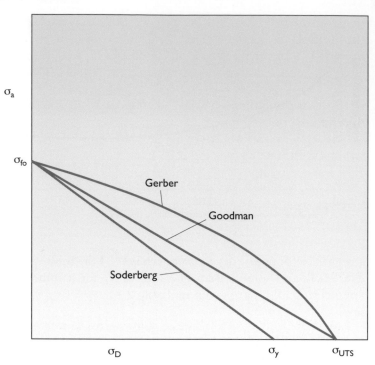

Fig. 2.15 Models relating mean stress to fatigue stress range.

the mean stress increases to the UTS, i.e.:

$$\sigma_{fm} = (1 - \sigma_m/\text{UTS}) \qquad (2.17)$$

The Goodman relation matches experimental observation quite closely for brittle metals, whereas the Gerber model, which matches experimental observations for ductile alloys, predicts a parabolic decline in fatigue stress with increasing mean stress:

$$\sigma_{fm} = \{1 - (\sigma_m/\text{UTS})^2\} \qquad (2.18)$$

The Soderberg model, which provides a conservative estimate of fatigue life for most engineering alloys, predicts a linear decrease in the fatigue stress with increasing mean stress up to σ_y:

$$\sigma_{fm} = (1 - \sigma_m/\sigma_y) \qquad (2.19)$$

Low-Cycle Fatigue

The stresses associated with low cycle fatigue are generally high enough to cause appreciable plastic deformation prior to fracture, and in these circumstances the fatigue life is characterised in terms of the *strain range*. Coffin and Manson noted that when the logarithm of the plastic strain amplitude, $\Delta\epsilon_p/2$, was plotted against the logarithm of the number of load reversals to failure, $2N_f$, for metallic materials, a linear result was obtained, i.e.

$$\Delta\epsilon_p/2 = \epsilon'_f(2N_f)^c \qquad (2.20)$$

where ϵ'_f is the fatigue ductility coefficient and c the fatigue ductility exponent.

The *total* strain amplitude $\Delta\epsilon/2$ is the sum of the elastic strain amplitude, $\Delta\epsilon_e/2$, and $\Delta\epsilon_p/2$.

but $\Delta\epsilon_e/2 = \Delta\sigma/2E = \sigma_a/E$, where E is Young's modulus.

so using equation 2.16 we obtain:

$$\Delta\epsilon_e/2 = \sigma'_f/E(2N_f)^b \qquad (2.21)$$

Combining equations 2.20 and 2.21, we can write the total strain amplitude:

$$\Delta\epsilon/2 = \sigma'_f/E(2N_f)^b + \epsilon'_f(2N_f)^c \qquad (2.22)$$

Equation 2.22 forms the basis for the strain-life approach to fatigue design, and Table 2.1 gives some strain-life data for some common engineering alloys.

Fatigue Crack Growth – The Use of Fracture Mechanics

The total fatigue life discussed so far is composed of both the crack nucleation and crack propagation stages. *Defect-tolerant* design, however, is based on the premise that engineering structures contain flaws, and that the life of the component is the number of cycles to propagate the dominant flaw – taken to be the largest undetectable crack size appropriate to the particular method of non-destructive testing employed.

Fracture mechanics may be employed to express the influence of stress, crack length and geometrical conditions upon the rate of fatigue crack propagation. Thus by employing equation (2.13a) a stress range $\Delta\sigma$ applied across a surface crack of length a, will exert a *stress intensity range*, ΔK, given by:

$$\Delta K = Y\Delta\sigma(\pi a)^{\frac{1}{2}} \qquad (2.23)$$

where Y is the geometrical factor.

Pre-cracked test-pieces may thus be tested under fluctuating stress, and the growth of the crack continuously monitored, for example by recording the change in resistivity of the specimen. After calibration, the resistivity changes may be interpreted in terms of changes in crack length, and the data may be plotted in the form of crack growth per cycle (da/

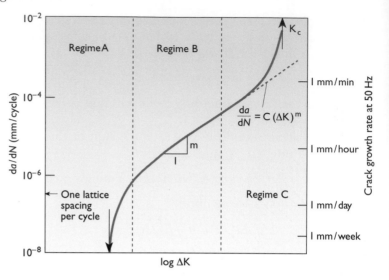

Fig. 2.16 Schematic illustration of fatigue crack growth curve.

dN) versus ΔK curves. Figure 2.16 illustrates schematically the form of crack growth curve obtained in the case of ductile solids.

For most engineering alloys, the curve is seen to be essentially sigmoidal in form. Over the central, linear, portion (regime B in Fig. 2.16) the fatigue crack growth rate (FCGR) is observed to obey the Paris power law relationship:

$$da/dN = C(\Delta K)^m \qquad (2.24)$$

where C and m are constants.

For tensile fatigue, ΔK refers to the range of mode I stress intensity factors in the stress cycle, i.e. $\Delta K = K_{\max} - K_{\min}$, and $K_{\min}/K_{\max}$ is known as R, the *load ratio*.

In the Paris regime, the FCGR is in general insensitive to microstructure of the material and to the value of R.

In regime A, the average FCGR becomes smaller than a lattice spacing, and this suggests the existence of a threshold stress intensity factor range, ΔK_{o}, below which the crack remains essentially dormant. The value of ΔK_{o} is not a material constant, however, but is highly sensitive to microstructure and also to the value of R, as apparent in the data for a variety of engineering alloys shown in Fig. 2.17: at high mean stress (high R), lower thresholds are encountered.

In regime C, FCG rates increase rapidly to final fracture. This corresponds to $K_{\max}$ in the fatigue cycle achieving the fracture toughness, K_c, of the material. As indicated in Fig. 2.16, this regime is again sensitive to the value of R.

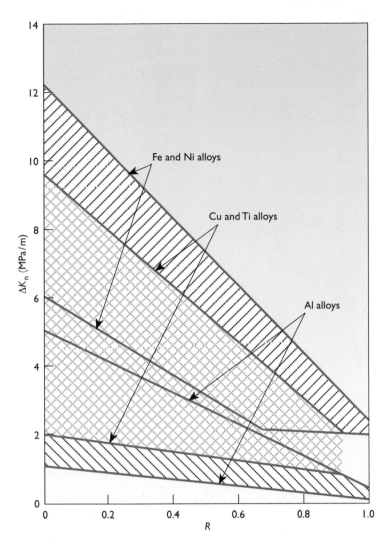

Fe and Ni alloys

Cu and Ti alloys

Al alloys

Fig. 2.17 Ranges of threshold ΔK versus load ratio (R) for some engineering alloys.

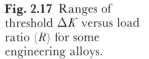

Fatigue Charts

Fleck, Kang and Ashby: *Acta Metall. Mater.* 1994, **42**, 365–381 have constructed some Material Property Charts for fatigue analogous to that illustrated in Fig. 1.1 which relates modulus to density of engineering materials. They are useful in showing fundamental realtionships between fatigue and static properties, and in selecting materials for design against fatigue. Thus Fig. 2.18 shows the well-known fact that the endurance limit σ_e scales in a roughly linear way with the yield stress, σ_y. The *fatigue ratio* defined as σ_e/σ_y at $R = -1$, appears as a set of diagonal lines. The ratio is near 1 for engineering ceramics, about 0.5 for metals and elastomers and about 0.3 for polymers, foams and wood.

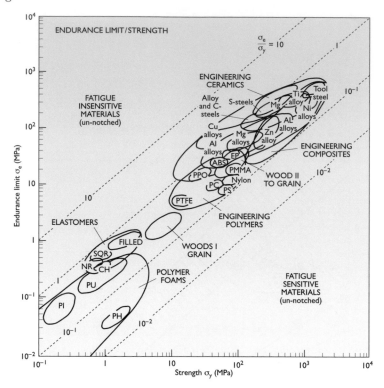

Fig. 2.18 Endurance limit versus yield strength for engineering materials.

Figure 2.19 refers to the behaviour of cracked specimens, and charts the relationship between the fatigue threshold and the fracture toughness of materials. The correlation between ΔK_{th} and K_{Ic} is less good than between σ_e and σ_y, reflecting the fact that, with the exception of polymers and woods, cracked materials are more sensitive to fatigue loading than those which are initially uncracked. The ratios of Figs 2.18 and 2.19 vary widely from material class to class, which emphasises the importance of employing fatigue properties rather than monotonic properties in design for cyclic loading.

Environment-Assisted Cracking

Materials are known to fail under stress when the *initial* stress intensity level is considerably below K_{Ic}. This is because cracks are able to grow to a critical size until the stress intensity level is increased to the critical value. There are a number of processes which may give rise to such crack extension in the presence of an applied static stress in an aggressive environment, namely **stress-corrosion cracking (SCC)**, **hydrogen embrittlement (HE)** and **liquid metal embrittlement (LME)**. These phenomena are referred to collectively as **environment-assisted cracking (EAC)**.

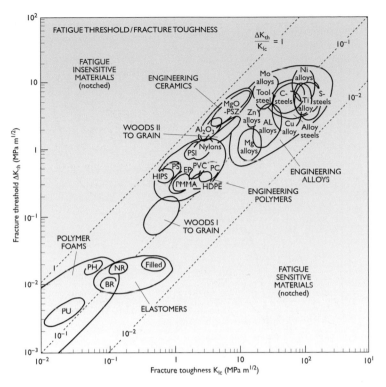

Fig. 2.19 Threshold stress intensity range versus fracture toughness of engineering materials.

da/dt – K Curves

The effect is encountered in polymers, ceramics and metallic materials. In order to quantify EAC processes, the rate of crack advance (da/dt) is monitored as a function of the instantaneous value of the stress intensity. Cracked specimens are tested in Mode I (crack opening mode), and in general the data are represented on plots of log da/dt versus K, and they take the form shown schematically in Fig. 2.20.

Three distinct crack growth regimes are shown (though all may not be present in a given material). In region I da/dt is strongly dependent on K, and in some materials the slope of this part of the curve is so steep that a value $K_{I\,EAC}$ may be defined, i.e. a level of K_I below which da/dt becomes vanishingly small.

In region II the rate of crack growth appears to be relatively independent of the prevailing level of K. The crack extension rate is still strongly dependent on the temperature, pressure and the environment, however. Finally, in region III da/dt is strongly K-dependent, and in the limit the crack grows unstably as K approaches K_{Ic}.

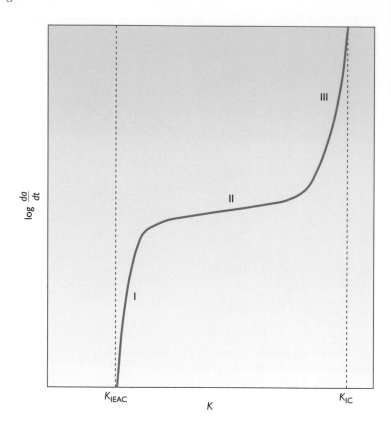

Fig. 2.20 Diagram showing three stages of environment-assisted cracking under sustained loading in an aggressive atmosphere. Lower and upper K limits of plot determined by K_{IEAC} and K_{IC}, respectively.

Time-Dependent Elastic Properties

Vibrating Solids

If a material is loaded elastically to a stress σ, the stored elastic energy per unit volume is:

$$U = \int_0^{\sigma_{max}} \sigma de = \frac{1}{2}\frac{\sigma^2}{E}$$

If it is loaded and then unloaded it dissipates an energy

$$\Delta U = \oint \sigma \, de$$

The *specific damping capacity*, D, is defined by the ratio:

$$D = \Delta U / U \qquad (2.25)$$

Another measure of damping is the *logarithmic decrement*, δ, which is the log of the ratio of successive amplitudes of natural vibration of the solid. The effect arises because the elastic forces acting may also assist thermal energy fluctuations to bring about atom movements, which are also time-dependent. The effect of these movements is to bring about time-dependent relaxations of stress in the lattice, an effect known as *anelasticity*, and each relaxation process has its own characteristic *time constant*. If, therefore, the damping is measured as a function of frequency of cyclic loading, a spectrum may be obtained with damping peaks occurring at characteristics frequencies, giving information about the molecular or atomic processes causing the loss or energy absorbing peaks.

The value of δ is very easily measured, for example by employing the specimen as a torsion pendulum. The logarithmic decrement usually has a very low value in metals, but it can rise to high values in polymers, with peaks occurring in the region of the so-called **glass-transition temperature**. In this temperature range, long-range molecular motion is hindered and damping is great.

Viscoelastic Behaviour of Polymers

We have seen in Chapter 1 that a thermoplastic polymer consists of a random mass of molecular chains, retained by the presence of secondary bonds between the chains. Under load, the chains are stretched and there is a continuous process of breaking and remaking of the secondary bonds as the polymer seeks to relax the applied stress. At low strains this anelastic behaviour is reversible in many plastics, and the laws of *linear viscoelasticity* are obeyed.

The linear viscoelastic response of polymeric solids has been described by a number of mechanical models which provide a useful physical picture of time-dependent deformation. These models consist of combinations of *springs* and *dashpots*. A spring element describes linear elastic behaviour:

$$\epsilon = \sigma/E \quad \text{and} \quad \gamma = \tau/G$$

where γ is the shear strain, τ the shear stress and G the shear modulus. A dashpot consists of a piston moving in a cylinder of viscous fluid, and it describes viscous flow:

$$\dot{\epsilon} = \sigma/\eta \quad \text{and} \quad \dot{\gamma} = \tau/\eta$$

where $\dot{\epsilon}$ and $\dot{\gamma}$ are the tensile and shear strain rates and η is the fluid viscosity which varies with temperature according to:

$$\eta = A\exp(\Delta H/RT)$$

where A is a constant, and ΔH is the viscous flow activation energy.

The Maxwell Model

If the spring and dashpot are in series (Fig. 2.21a), the stress on each is the same and the total strain or strain rate is determined from the sum of the two components:

$$\frac{de}{dt} = \frac{\sigma}{\eta} + \frac{1}{E}\frac{d\sigma}{dt}$$

The Voigt Model

When they are in parallel (Fig. 2.21b), the strains in the two elements are equal, and total stress on them is given by the sum:

$$\sigma_T(t) = Ee + \eta\frac{de}{dt}$$

if the applied stress is constant, as in a creep test, this differential equation may be directly integrated to give the strain as a function of time:

$$e(t) = \frac{\sigma}{E}(1 - e^{-t/\tau})$$

where the time constant, $\tau = \eta/E$.

The Four Element Model

A more realistic description of polymer behaviour is obtained with a four element model consisting of Maxwell and Voigt elements in series (Fig. 2.21c). By combining the above equations, the total strain experienced by this model is given by:

$$e(t) = \frac{\sigma}{E_1} + \frac{\sigma}{E_2}(1 - e^{-t/\tau}) + \frac{\sigma}{\eta}t$$

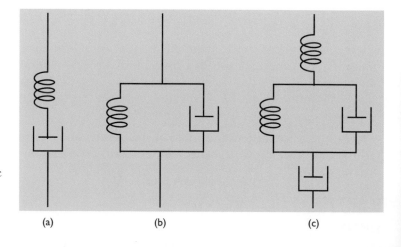

Fig. 2.21 Spring-and-dashpot models of viscoelastic solids: (a) Maxwell, (b) Voigt and (c) four-element model.

(a)   (b)   (c)

It should be borne in mind that many additional elements are often required to represent adequately the mechanical behaviour of a given engineering polymer.

Having reviewed the **structure** of engineering materials, and considered how their **mechanical properties** may be defined and measured, the remaining four chapters will explore the relationship between structure and properties, taking each family of materials in turn.

PART II: STRUCTURE/PROPERTY RELATIONSHIPS

3 Metals and Alloys

GENERAL STRENGTHENING MECHANISMS: THE EFFECT OF PROCESSING

The strength of a crystal may be assessed either by its resistance to *cleavage*, by the application of normal stresses across the cleavage plane, or by its resistance to *shear* under the action of shear stresses on the slip plane. It is possible to make an estimate of these strengths for ideal, or perfect crystals, but in this case a large discrepancy is found between these values and those measured experimentally. In pure materials, real strengths are several orders of magnitude lower than their theoretical strengths – cleavage occurs due to the presence of *cracks* in brittle solids and shear occurs due to the presence of mobile *dislocations* in ductile solids. The local displacement (b) associated with a given dislocation is known as its *Burgers vector*.

Dislocations are linear defects which are always present in technical materials, usually in the form of a three-dimensional network. Dislocation density (ρ) is usually expressed as the length of dislocation line per unit volume of crystal, or its geometrical equivalent – the number of dislocations intersecting unit area. In a reasonably perfect crystal the value of ρ might be of the order 10^9 m^{-2}. Their number per unit *length* will therefore be $\rho^{\frac{1}{2}}$, and their average *separation* will be the reciprocal of this quantity, giving ~ 30 μm for the dislocation density considered.

Work Hardening

As a crystal is deformed, dislocations multiply and the dislocation density rises. The dislocations interact elastically with each other and the average spacing of the dislocation network decreases. The shear yield strength (τ) of a crystal containing a network of dislocations of density ρ is given by:

$$\tau = \alpha \, Gb(\rho^{\frac{1}{2}}) \tag{3.1}$$

where G is the shear modulus, b the dislocation Burgers vector and α is a constant of value about 0.2. As plastic deformation continues, therefore, the increase in dislocation density causes an increase in τ – the well-known effect of *work hardening*, Fig. 2.2.

As a technical means of producing a strong material, work hardening can only be employed in situations where large deformations are involved, such as in wire drawing and in the cold-rolling of sheet. This form of hardening is lost if the material is heated, because the additional thermal energy allows the dislocations to rearrange themselves, relaxing their stress fields through processes of **recovery** and being annihilated by **recrystallisation** (see below).

Grain Size Strengthening

Metals are usually used in polycrystalline form, and in this situation dislocations are unable to move long distances without being held up at grain boundaries. Metal grains are not uniform in shape and size – their three-dimensional structure resembles that of a soap froth. There are two main methods of measuring the grain size:

(a) The mean linear intercept method which defines the average chord length intersected by the grains on a random straight line in the planar polished and etched section, and

(b) The ASTM comparative method, in which standard charts of an idealised network are compared with the microstructure. The ASTM grain size number $(\mathcal{N})$ is related to n, the number of grains per square inch in the microsection observed at a magnification of $100\times$, by:

$$\mathcal{N} = \frac{\log n}{\log 2} + 1.000$$

Thus the smaller the average grain diameter (d), the higher the ASTM grain size number, $\mathcal{N}$.

The tensile yield strength (σ_y) of polycrystals is higher the smaller the grain size, these parameters being related through the **Hall-Petch** equation:

$$\sigma_y = \sigma_0 + k_y \, d^{-\frac{1}{2}} \tag{3.2}$$

where k is a material constant and σ_0 is the yield stress of a single crystal of similar composition and dislocation density.

The control of grain-size in crystalline materials may be achieved in several ways:

From the Molten State

As we discussed in Chapter 1, when molten metal is cast, the final grain size depends upon the rate of nucleation of solid crystals within the melt. Equation 1.6 shows that the critical nucleus size decreases with the degree of supercooling of the liquid, so that rapidly cooled liquids will form solids of finer grain size than slowly cooled liquids. For example, when large bronze sand castings are made (e.g. a ship's propeller), the cooling rate and degree of supercooling of the melt are low, so the product has a coarse grain-size – typically tens of mm in dimensions. Conversely, when die-cast objects are formed by injecting molten metal into a water-cooled metal die, the high supercooling leads to a high nucleation-rate and thus a fine grain-size – typically tens of μm in dimensions.

Fine-grained castings can be produced if an *inoculant* is added to the molten metal before it is introduced into the mould. For example, the addition of a small quantity of zirconium to molten magnesium alloys results in a dramatic refinement of the grain size in the casting. In recent years, ultrafine (sub-micrometre) grain-size material has been produced by *rapid-solidification processing* (RSP), whereby molten metal is sprayed in fine droplet form on to a water-cooled substrate, thus achieving extremely high supercoolings. The high cost of this process limits its wide application, however.

Solid State Phase Changes

In the case of those materials which undergo a phase change in the solid state, it is possible to refine the grain size by thermal treatment. The most important example of this process is that of *normalising* steel, whereby a fine-grained microstructure can be developed in (say) a coarse-grained steel casting by subjecting it to an appropriate heating and cooling cycle.

Pure iron (α-iron, or 'ferrite') undergoes a change in crystal structure when heated above 910°C, forming γ-iron, or 'austenite'. Most ferritic steels exhibit a similar transformation, and if such a steel is heated just above its γ transformation temperature, a new (γ) grain structure will form. When the material is allowed to undergo normal air-cooling, the γ will transform to new α-grains by a process of nucleation and growth. Air-cooling permits sufficient supercooling to encourage prolific nucleation of α, so the original grain structure is refined. If the normalising cycle is repeated, an even finer grain-size can be achieved. This effect is illustrated in Fig. 3.1 for a steel specimen of initial grain size 50 μm (ASTM 6). If the specimen is immersed in a molten lead bath at 815°C and then allowed to air cool to room temperature, a structure of grain size 11 μm (ASTM 10) is formed. If the cycle is repeated grain sizes of 8 μm (ASTM 11) and 5 μm (ASTM 12) are formed.

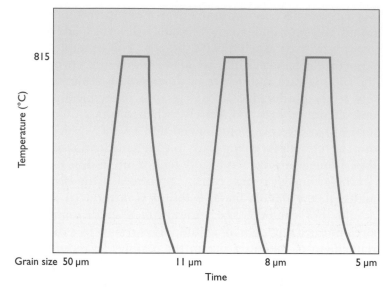

Fig. 3.1 Change grain size with normalising for a ferritic steel.

Recrystallisation

If work-hardened metals are annealed at a suitable temperature, a set of new, undeformed grains will grow by a process of nucleation and growth – consuming the so-called cold-worked microstructure. The process, known as *recrystallisation*, is illustrated in Fig. 3.2. If the temperature is raised, or the annealing prolonged, these new grains grow at the expense of their neighbours – a phenomenon known as *grain growth*.

A minimum degree of prior cold work is necessary before a

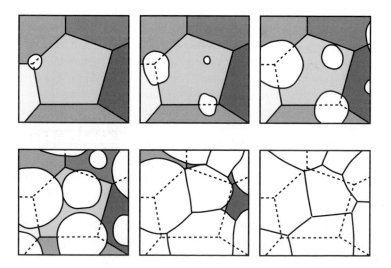

Fig. 3.2 Recrystallisation: new strain-free grains progressively replacing a cold-worked structure.

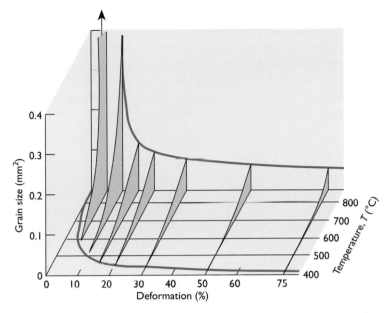

Fig. 3.3 Showing the recrystallised grain size as a function of prior deformation and recrystallisation temperature.

material will recrystallise, and the minimum temperature for recrystallisation (T_R) is dependent on several factors:

(i) it is inversely proportional to the *time* of anneal, and to the degree of prior cold work.

(ii) it is directly proportional to the initial grain-size, and to the temperature of prior cold work.

The effect of the degree of prior strain and the temperature of anneal upon the final grain size produced after a recrystallisation is indicated in Fig. 3.3, where is it seen that in general coarse grain sizes result after small strains and high annealing temperatures, whereas fine grained structures form after high strains and low annealing temperatures.

If the degree of prior strain or the annealing temperature is too low to cause recrystallisation, the material may still undergo softening by a process known as *recovery*. Here no microstructural change is apparent, but some lattice defects are removed by the thermal treatment.

Thermo-Mechanical Treatment

When materials are *hot-worked*, *dynamic recrystallisation* may take place during the deformation process itself. Depending on the degree of strain and the temperature of working, therefore, the grain size of the final product may be controlled.

Hot working a material which may undergo a phase change on cooling, such as steel, presents a further, powerful means of grain size control. *Controlled rolling* of steel is an example of this, whereby the steel is deformed above the γ transformation temperature: dynamic recrystallisation produces a fine γ grain size, which, on air-cooling, is transformed to an even finer α grain size. Sophisticated process control is necessary to produce material consistently with the desired microstructure, but, in principle, controlled rolling constitutes a very attractive means of achieving this.

Alloy Hardening

Work hardening and grain-size strengthening, which we have considered so far, can be applied to a pure metal. The possibility of changing the composition of the material by alloying presents further means of strengthening. We will consider two ways in which alloying elements may be used to produce strong materials:

Solute Hardening

We have shown in Figs 1.10a and 1.10b that two types of solid solution may be formed, namely *interstitial* and *substitutional* solutions. The presence of a 'foreign' atom in the lattice will give rise to local stresses which will impede the movement of dislocations, hence raising the yield stress of the solid.

This effect is known as *solute hardening*, and its magnitude will depend on the concentration of solute atoms in the alloy and also upon the magnitude of the local misfit strains associated with the individual solute atoms. It is also recognised that the *solubility* of an element in a given crystal is itself dependent upon the degree of misfit – indeed if the atomic sizes of the solute and solvent differ by more than about 14%, then only very limited solid solubility occurs. There must thus be a compromise between these two effects in a successful solution-hardened material – i.e. there must be sufficient atomic misfit to give rise to local lattice strains, but there must also be appreciable solubility.

A theoretical approach expresses the increase in shear yield stress, $\Delta\tau_y$, in terms of a solute atom mismatch parameter, ϵ, in the form:

$$\Delta\tau_y = G\epsilon^{3/2} b\, c^{\frac{1}{2}} \tag{3.3}$$

where G is the shear modulus, b the dislocation Burgers vector and c the concentration of solute atoms.

Precipitation Hardening

Thermal treatment can be used to control the size and distribution of second-phase particles in any alloy which undergoes a phase transformation in the solid state. In many alloy systems, the solid solubility changes with temperature in the way illustrated in Fig. 1.19. Above temperature T_1, a single phase (α) solid solution exists, and if the material is quenched rapidly from this temperature range, a supersaturated α solid solution is formed – the degree of supersaturation increasing with increasing supercooling below the solvus line, ab. If the temperature is then raised again in order to allow solid state diffusion to proceed, the supersaturation will be relieved by the nucleation and growth of a precipitated second phase.

In alloys of relatively low melting-point (in aluminium alloys, for example) there will be an appreciable diffusion rate of solute atoms at room temperature, so that over a period of time, a second phase will precipitate out in a very finely-divided form. This effect is known as 'ageing', but in most alloys the temperature has to be raised in order to cause precipitation to occur and the material is said to be 'artificially aged'. The ageing temperature affects the precipitate size in the manner illustrated schematically in Fig. 3.4. Low ageing temperatures correspond to high supersaturation, and prolific nucleation of precipitates occurs, whereas at higher

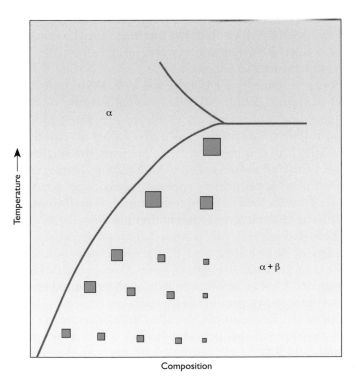

Fig. 3.4 Variation of precipitate size with ageing temperature.

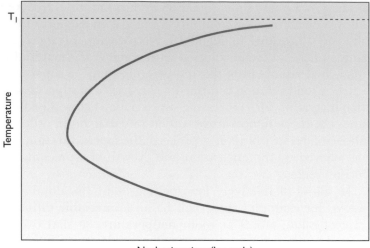

T_1

Temperature

Nucleation time (log scale)

Fig. 3.5 A TTT diagram.

ageing temperatures (lower supersaturation) fewer, coarser particles are formed. The **kinetics** of precipitation can be represented by a **temperature-time-transformation (TTT)** diagram, Fig. 3.5. At small undercoolings, there is a long incubation period, due to the low probability of formation of the (large) critical nucleus (equation 1.6). As the supercooling increases, the nucleation rate will increase, since the critical nucleus size is smaller. The lower the transformation temperature, therefore, the more prolific the nucleation and the finer the dispersion of particles. However, the lower the temperature the more sluggish the solid state diffusion will become, and the TTT curve has a 'C' shape indicating a more sluggish transformation at low temperature.

Quenching and ageing are therefore very powerful means of controlling the distribution of a precipitate of second phase in an alloy. After quenching the alloy from the single-phase region of the phase diagram, a high ageing temperature is selected if a coarse, widely spaced dispersion of particles is required, and a lower ageing temperature is used to produce the second phase is a more finely divided form.

These precipitates can have a profound effect upon the mobility of dislocations, and it is possible to produce large changes in the yield strength of such alloys by suitable heat-treatment. A great advantage is that the required strength can be induced in a product at the most convenient stage in its manufacture. For example, the alloy may be retained in a soft form throughout the period when it is being shaped by forging, and it is finally hardened by precipitation in order to give it good strength in service.

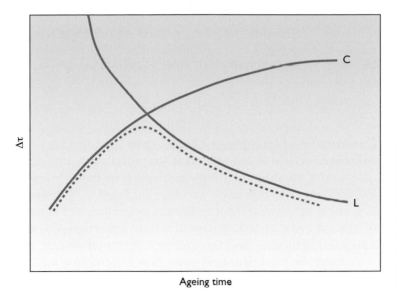

$\Delta\tau$

Ageing time

Fig. 3.6 Showing change in yield stress ($\Delta\tau$) with ageing time for a precipitation-hardening alloy. Curve C is followed if the precipitates are cut by dislocations, and curve L is followed if the dislocations loop between the particles. The response is given by the dashed curve.

On being held up by a precipitate, a dislocation can continue in its path across the crystal in two possible ways. If the particles are very close together, the dislocation may *cut through* each particle, but if the particles are further apart, the dislocation may *loop between* the particles. During the ageing process, as the particles grow the stress increment required to make the dislocations cut them also rises (Fig. 3.6, curve *C*). The increase in shear stress due to precipitate cutting, $\Delta\tau_c$, is given by an equation of the form:

$$\Delta\tau_c = Af^{\frac{1}{2}}r^{\frac{1}{2}} \qquad (3.4)$$

where r is the particle radius, f the volume fraction of precipitate, and A a material constant.

As ageing proceeds, the particles gradually increase in size and the average spacing between the particles also increases. The stress increment to cause dislocation looping ($\Delta\tau_l$) decreases as the inter-particle spacing increases (curve *L*, Fig. 3.6), depending on the precipitate size and volume fraction according to:

$$\Delta\tau_l = B.Gbf^{\frac{1}{2}}r^{-1} \qquad (3.5)$$

where B is a constant dependent upon precipitate particle shape.

As ageing continues, the measured yield stress would therefore be expected to follow the form of the dotted curve in Fig. 3.6, and this general pattern of behaviour with an optimum ageing time to give a maximum hardness is commonly observed in many commercial alloys. The time to peak hardness depend on the solute diffusion rate, and thus on the ageing temperature.

Combinations of Strengthening Mechanisms

Most commercial alloys owe their strength to a combination of several of the strengthening mechanisms we have reviewed. We will return to this later, and Figs 3.22 and 3.26 illustrate how the strength of certain steels may be understood in terms of the additive contributions of grain size strengthening, solute strengthening and the presence of second phases.

Perhaps the most dramatic example of strengthening from several mechanisms is the formation of *martensite* when steel is rapidly quenched from a high temperature (see Ch.1). An extremely hard (and brittle) phase is formed in this diffusion-less transformation. Martensite owes its strength to the combination of a high dislocation density, a very fine grain size and a high supersaturation of solute atoms (carbon).

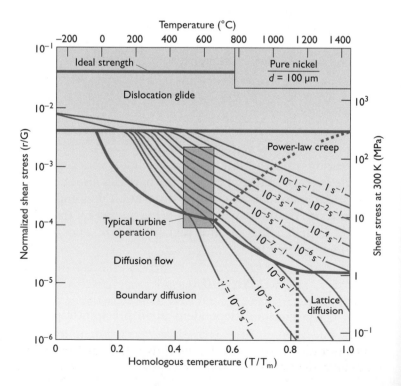

Fig. 3.7 The deformation mechanism map for nickel of grain size 100 μm.

Strength at High Temperature – Creep-Resistant Alloys

A prerequisite for an engineering component operating at elevated temperatures is that it should be resistant to degradation by oxidation and corrosion from its environment. This aspect of material behaviour will be considered in more detail later (see pp. 118–121).

At low temperatures materials deform by the glide of dislocations, and the principle of designing strong materials is essentially one of introducing barriers to this process. As the operating temperature increases other mechanisms of flow become possible – dislocations can *climb* over barriers, grains slide over each other at grain boundaries, vacancies diffuse. Thus in a polycrystalline solid there are many distinguishable mechanisms by which it can flow. In one range of stress and temperature one of these flow mechanisms is dominant, in another range a different mechanism will obtain

This 'landscape' of deformation mechanisms is most conveniently surveyed with the aid of a 'deformation mechanism map', which summarises, for a given polycrystalline solid, information about the range of dominance of each of the mechanisms of plasticity and the rates of flow they produce. Figure 3.7 is such a map for pure nickel with a grain size of 100 μm, the coordinate axes are temperature, T, normalised in respect to the melting point, T_m, and normalised shear stress, τ/G, on a logarithmic scale.

At low temperatures flow is confined to the *dislocation glide* field, slip being the dominant mechanism. Above 0.3 T_m dislocations can climb and the shear strain rate ($\dot{\gamma}$) can be characterised by an equation of the form:

$$\dot{\gamma} \propto (\tau/G)^n \tag{3.6}$$

This regime is called *power-law creep*.

If the stress is too low to permit dislocation movement, the material may still undergo strain by *diffusional flow* of single ions at a rate which depends strongly on the grain size (d):

$$\dot{\gamma} \propto \tau/d^m \tag{3.7}$$

In this regime two subgroups exist, corresponding to diffusion occurring predominantly through the grains ($m = 2$) (Nabarro-Herring creep) or round their boundaries ($m = 3$) (Coble creep).

Intrinsically high creep strength is expected in materials in which diffusion is slow, i.e. in materials of high melting-point and in covalent elements and compounds. Further reduction in power-law creep may be achieved by metallurgical control of the microstructure: solute and particle hardening, together with a grain-boundary precipitate to suppress grain-boundary sliding are all commonly encountered in creep resistant materials.

The most sophisticated family of alloys exemplifying these principles are the age-hardening nickel-based 'superalloys' which are widely used in gas turbines. If power-law creep is effectively suppressed by these means, diffusion creep will be reduced by increasing the grain-size (equation 3.7), and the use of single crystal turbine blades is an extreme example of this approach.

Resistance to Fatigue Failure

It has long been recognised that fatigue cracks normally nucleate at a free surface in a component, and often at a point of local stress concentration. There are thus two approaches to the avoidance of fatigue, namely careful *design* in order to avoid concentrations of tensile stresses in the surface of the component and secondly by *microstructural control*.

Fatigue-Resistant Design

In component subject to fluctuating stresses, it is essential to avoid sharp changes in cross-section. Notches and grooves, such as keyways on rotating shafts are obvious examples of such features, and smooth, gradual changes in cross-section, with large radii of curvature at fillets are essential if early fatigue fracture is to be avoided. Figure 3.8 shows a series of test-pieces prepared from identical 10 mm diameter steel bars. In each case the minimum cross-section of the test-piece was 7 mm, and various means of effecting the change in cross-section from 10 mm to 7 mm have been employed. S–N curves were constructed from the results of fatigue tests, and the endurance limits (S) determined for each specimen design.

It is seen that identical values of endurance limit were obtained for specimens with fillet radii of 250 mm and 25 mm. When the fillet radius was reduced to 6.5 mm, however, it is seen that a significant drop in endurance limit took place, due to the concentration of stress in the surface associated with the sharp change in cross-section. In specimens containing a sudden step or a sharp notch, the magnitude of the endurance limit decreases further, being in the notched specimen less than half of that measured in the smooth specimens.

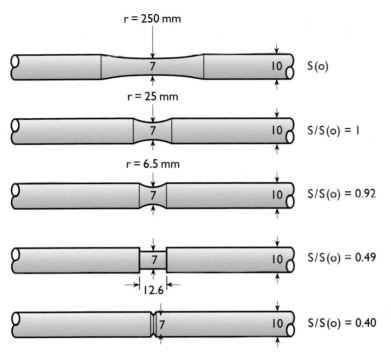

Fig. 3.8 Effects of fillet radii and notches on fatigue limit (S).

Fatigue-Resistant Microstructures

Fatigue cracks may nucleate at stress concentrations at weak *internal* surfaces and interfaces in a material, as well as at the external surface. Internal voids and cavities are therefore undesirable, and such features as blowholes and shrinkage cavities in castings, as well as imperfect welds in fabricated components are obviously deleterious. Careful **non-destructive testing** is often applied to critical components in order to minimise the danger of fatigue failure originating from internal defects of this sort.

Coarse inclusions of a second phase can also provide a source of local internal stress, leading to the early nucleation of fatigue cracks . Coarse graphite inclusions are present in many cast irons, and are the origin of their poor fatigue resistance. Inclusions of slag etc. may be present in steel, and sophisticated techniques such as vacuum melting and degassing are employed to produce 'clean steels' with enhanced properties. Coarse inclusions of intermetallic phases may be present in a wide range of alloys if special precautions are not taken; for example a low iron content is specified in aluminium alloys for aerospace applications in order to minimise the occurrence of such particles.

Surface treatment of a component is a widely employed method of enhancing resistance to fatigue failure, and a number of approaches are possible. Clearly, *surface finish* is important, and the removal of scratches and machining marks can be significant in this context, with surface polishing being a further possibility. Secondly, *surface hardening* can be employed to inhibit fatigue; this can be effected in several ways:

(a) Work hardening a shallow surface layer can be achieved by 'shot blasting' or 'shot peening'. Surface rolling also enhances fatigue resistance for the same reason. The beneficial effect will be reduced, of course, if the treatment introduces any roughening of the surface.

(b) In steels, surface hardening can be achieved by changing the surface composition by diffusion heat-treatment of the component. *Nitriding* is an important example of this approach. The machine-finished components are produced from a steel containing about 1% aluminium, and then nitrogen is allowed to diffuse into the surface at about 500°C. A hard, shallow layer is formed, whose depth (usually less than 1 mm) is dependant on the time of treatment. The layer contains fine particles of aluminium nitride, whose formation introduces highly localised surface compressive stresses, resulting in significant improvement in fatigue resistance.

THE FAMILIES OF ENGINEERING ALLOYS

We will now survey the structure property relationships of the important engineering alloys. We will consider them in three groups, namely:

1. The light alloys, based on aluminium, magnesium and titanium.
2. The 'heavy' non-ferrous alloys based on copper, lead, zinc and nickel.
3. The ferrous alloys, namely steels and cast irons.

A reading list is included at the end of this chapter, suggesting books giving fuller details of the physical metallurgy of the various families of alloys. Tables are included which surveys the nomenclature of industrially important alloys, and outlines the designations employed to describe their metallurgical condition, or *temper*.

Aluminium Alloys

Cast Aluminium Alloys

About 20% of the world production of aluminium is used for cast products. Aluminium alloys have a relatively low melting temperature, but exhibit a high *shrinkage* during solidification. Shrinkage of between 3.5 and 8.5% may occur, and allowance has to be made for this in mould design in order to achieve dimensional accuracy in the product.

Aluminium–silicon alloys are the most important of the aluminium casting alloys, and the relevant phase diagram is shown in Fig. 3.9, which is seen to be of simple eutectic form. Slow solidification produces a very coarse eutectic structure consisting of large plates or needles of silicon in an aluminium matrix. The silicon plates are brittle, so castings with this coarse eutectic exhibit low ductility. Refinement of the eutectic improves the mechanical properties of the casting, and this can be brought about rapid cooling, as in permanent mould casting, or by *modification*, which involves adding sodium salts or metallic sodium to the melt prior to pouring. Improved tensile strength *and* improved ductility is

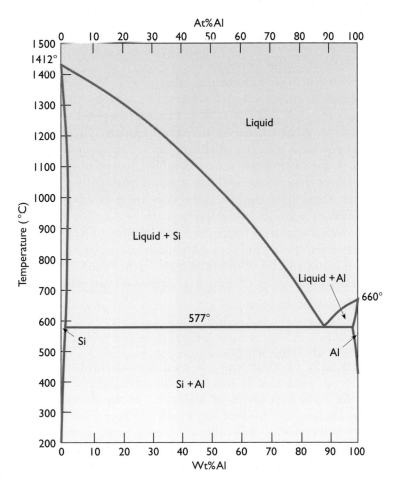

Fig. 3.9 The Silicon–Aluminium phase diagram.

the result of this microstructural change – a combination which is rarely encountered in physical metallurgy.

If the silicon content is below 8% modification is unnecessary, because the primary aluminium phase is present in sufficient quantity to confer adequate ductility. The binary Al–Si alloys are generally used where strength is not a primary consideration, e.g. pump casings and automobile engine manifolds. Copper and magnesium additions are made to enhance the strength of such alloys, and more complex compositions (e.g. including nickel additions) lead to improved elevated temperature properties for such applications as piston alloys for internal combustion engines.

Small additions of magnesium allows significant age-hardening of the castings through precipitation of Mg_2Si in the aluminium matrix. Doubling of the yield strength may be achieved in this way, and such alloys find use in aircraft and automotive applications.

Aluminium–copper alloys, although less easily cast than Al–Si alloys, respond well to age-hardening heat treatments. Several compositions have been developed with enhanced elevated temperature properties, for example for use as diesel engine pistons.

Aluminium–magnesium alloys, with Mg contents in the range 4–10% , are characterised by a high resistance to corrosion. Only Al–10% Mg castings respond to heat-treatment. **Aluminium–zinc–magnesium** alloys have a relatively high eutectic melting point, which make them suitable for castings that are assembled by brazing. The as-cast alloys respond to both natural and artificial ageing.

Cast aluminium and its alloys may be *grain refined* by the addition of suitable innoculants to the melt. For example commercial additives based on a master alloy of Al–Ti–B are widely used: these produce intermetallic particles in the melt which act as centres of crystallisation for the alloy. Table 3.1 presents some properties of a selection of cast aluminium alloys.

Wrought Aluminium Alloys

About 85% of aluminium is used for wrought products, produced from cast ingots by rolling, extrusion, drawing etc. An outline is given in Table 3.2 of the International Alloy Designation System employed for these materials.

We will consider these alloys in two groups, namely those whose properties are *not* enhanced by heat-treatment, and those that are.

Table 3.1 Typical mechanical properties of some cast aluminium alloys

Alloy	Condition	Density Mg m^{-3}	Young's Modulus GPa	Proof stress MPa	UTS MPa	Elongation (% in 50 mm)
Al-11.5Si (LM6)	Sand cast	2.65	71	65	170	8
Al-5Mg-0.5Mn (LM5)	Sand cast	2.65	71	100	160	6
Al-6Si-4Cu-0.2Mg-1Zn (LM21)	Sand cast	~2.7	71	130	180	1

Table 3.2 The International Alloy Designation System (**IADS**) for wrought aluminium alloys.

4-digit series (xxxx)

Each wrought alloy is assigned a four-digit number of which the first digit is determined by the major alloying element(s) present, thus:

Series	Main alloying elements
1xxx	Unalloyed aluminium (99% Al minimum)
2xxx	copper
3xxx	manganese
4xxx	silicon
5xxx	magnesium
6xxx	Mg and Si
7xxx	zinc
8xxx	others

Temper or heat treatment (also applied to Mg alloys)

Suffix letters and digits are added to the alloy number in order to specify the mechanical properties of the alloy and the way in which the properties were achieved, thus:

Suffix letter	Basic condition
F	as-fabricated
O	annealed wrought products
H	cold worked (strain hardened)
T	heat treated

Suffix digits	
First digit	secondary treatment
Second digit (H only)	degree of cold work

Recourse to detailed specifications or to manufacturers' literature is suggested when several digits are included in the temper designation.

Non-Heat-Treatable Alloys

There are two important families:

Aluminum–manganese alloys (3xxx series) contain up to 1.25% Mn, which gives rise to solution-hardening. The further addition of magnesium gives a further increase in strength, coupled with high ductility and excellent corrosion resistance. These alloys are widely used for cooking utensils, as well as for beverage cans.

Aluminium–magnesium alloys (5xxx series) contain up to 5% Mg. The alloys owe their strength to work hardening, which occurs at a rate that increases as the Mg content is raised. Over a period of time the tensile properties may decline due to localised recovery, but special tempers may be applied which stabilise them against this effect.

The alloys are widely used in welded applications. Their corrosion resistance makes them suitable for storage tanks and for marine hulls and superstructures.

Heat-Treatable Alloys

There are three important families:

Aluminium–copper alloys (2xxx series). These can contain up to 6.8% copper, and this system has been most widely studied as an example of *age-hardening*. For example, alloy 2219 (6.3% Cu) is available as sheet, plate and extrusions, as well as forgings, and it can be readily welded. It has relatively high strength in the peak-aged condition, but its peak hardness may be enhanced by about one third by strain hardening the quenched, supersaturated alloy prior to artificial ageing.

The Al–Cu–Mg alloy known as Duralumin was the earliest age-hardening alloy to be developed, by Wilm in 1906. Alloy 2014 is a development of Duralumin, and is still widely used for aircraft construction. The addition of copper is deleterious to the corrosion resistance of aluminium, so that sheet material in this group of alloys are often roll clad with pure aluminium or Al–1% Zn to produce a corrosion-resistant surface layer. This soft surface layer can lead to serious deterioration of the fatigue resistance, however.

Aluminium–magnesium–silicon alloys (6xxx series) are medium-strength alloys, widely used in the form of extruded sections for structural and architectural application – window frames are a typical example. The alloy can be quenched as it emerges from the extrusion press, thus eliminating the need for a separate solution-treatment operation. Optimum properties are again developed by a final ageing treatment.

Table 3.3 Typical mechanical properties of some wrought aluminium alloys.

Alloy	Condition	Density Mg m^{-3}	Young's Modulus GPa	Proof stress MPa	UTS MPa	Elongation (% in 50 mm)
Al-4.5Cu-1.5Mg-0.6Mn (2024)	Peak aged	2.77	73	395	475	10
Al-1.25Mn (3103)	Annealed Work hardened	2.74	69	65 185	110 200	40 7
Al-2Mg-0.3Mn (5251)	Annealed Work hardened	2.69	70	60 215	180 270	20 4
Al-0.5Mg-0.5Si (6063)	Peak aged	2.70	71	210	245	20
Al-5.6Zn-2.5Mg-1.6Cu-0.25Cr (7075)	Peak aged	2.80	72	500	570	11
Al-2.5Li-1.3Cu-0.95Mg-0.1Zr (8090)	Peak aged	2.55	77	436	503	5

Aluminium–zinc–magnesium alloys (7xxx series) are the highest-strength family of aluminium alloys. They are readily welded, and find wide structural application. An addition of copper is made to reduce the susceptibility to stress-corrosion cracking (SCC), and the Al–Zn–Mg–Cu alloys are widely used in aircraft construction. Alloy 7075 is the most widely known of these, and often rather complex heat-treatments have to be applied in order to minimize the propensity to SCC.

A new family of age-hardening aluminium alloys containing **lithium** (8xxx series) has been developed in recent years which have the advantage of a lower density and a higher value of Young's modulus than the 7xxx series. They are designed to substitute for conventional aircraft alloys, with a density reduction of 10% and a stiffness increase of at least 10%. Their purchase price is several times that of existing high-strength aluminium alloys, so their overall economic advantages have to be very carefully weighed when considering a potential application.

Table 3.3 gives the mechanical properties of some wrought aluminium alloys.

Magnesium Alloys

Cast Magnesium Alloys

Up to 90% of magnesium alloys are produced as castings,

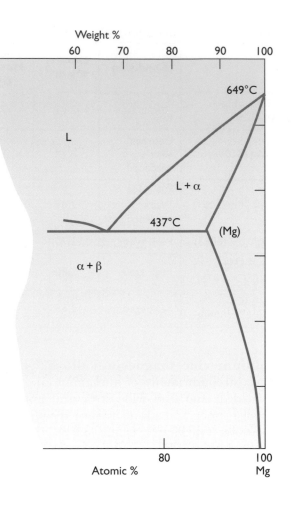

Fig. 3.10 The Mg-rich end of the Mg–Al phase diagram.

widely used in the aerospace industries. **Magnesium–aluminium** alloys contain 8–9% Al with up to 2% of zinc to increase the strength and 0.3% Mn which improves the corrosion resistance. From the Mg–Al phase diagram, Fig. 3.10, it will be seen that increasing quantities of the β-phase ($Mg_{17}Al_{12}$) will be formed as the Al-content increases above 2%. This is accompanied by an increase in proof stress and a decrease in % elongation in the material.

Grain refined castings are produced by adding **zirconium**, and a series of Mg–Zn–Zr alloys have been developed which also respond to age-hardening.

Wrought Magnesium Alloys

The most widely-used sheet alloy is Mg–3Al–1Zn–0.3Mn, which is strengthened by strain hardening. A range of extrusion alloys exist, based on Mg–Al–Zn, with Al contents

Table 3.4 Typical mechanical properties of some magnesium alloys.

Alloy	Condition	Density Mg m^{-3}	Young's Modulus GPa	Proof stress MPa	UTS MPa	Elongation (% in 50 mm)
Mg-6Al-3Zn-0.3Mn (AZ63)	Sand cast Peak aged	~1.8	~45	75 110	180 230	4 3
Mg-9.5Al-0.5Zn-0.3Mn (AZ91)	Cast and peak aged	~1.83	~45	127	239	2
Mg-6.5Al-1Zn-0.3Mn (AZ61)	Extruded	~1.8	~45	180	260	7

between 1 an 8%, but the highest room temperature strength is found in the Mg–6Zn–0.7Zr alloy when it is aged after extrusion.

The mechanical properties of some magnesium alloys are given in Table 3.4.

Titanium Alloys

Aerospace applications account for some 80% of titanium that is produced – most of the remainder being used in the chemical industry since its protective oxide film gives it relative chemical inertness. It shows an outstanding resistance to corrosion by body fluids which is superior to that of stainless steels, and this has led to its use for prosthetic devices. Its low density and high melting point (1678°C) has also led to the development of titanium alloys for certain critical gas turbine components.

At room temperature Ti has a hexagonal crystal structure (α) which changes to a body-centred cubic structure (β) at 882°C that remains up to the melting point. There are thus three types of Ti alloy microstructure, namely those with α, β and mixed α/β structures.

α-alloys

α-alloys can be divided into three subgroups (i) *single-phase α* which owe their strength to solute hardening, (ii) *near-α* alloys which contain up to 2% of some β-stabilizing elements and may thus be forged and/or heat-treated in other phase fields, and (iii) *alloys which respond to conventional age-hardening treatment*.

(i) The most widely used **single-phase** α material is in fact Commercially Pure (CP) titanium, which is essentially a Ti–O alloy, solution-hardened by controlled amount of added

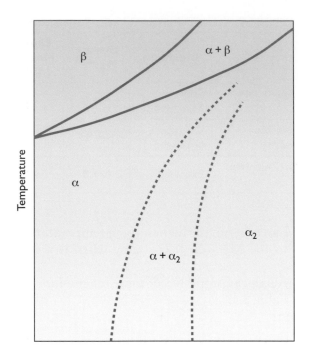

Fig. 3.11 Schematic phase diagram for near-α alloys of titanium.

oxygen, which dissolves interstitially in the metal. Higher strengths are achieved by substitutional solid solution hardening, of which Ti–5Al–2.5Sn is an example.

(ii) The **near-α** alloys show the greatest creep resistance above 400°C of all Ti alloys, and have attractive properties for use as a compressor disc alloy in gas turbines. A typical example is Ti–6Al–5Zr–0.5Mo–0.25Si, known as IMI 685.

Alloys in categories (i) and (ii) have phase diagrams of the form shown in Fig. 3.11.

(iii) Ti–2.5Cu may be strengthened by a classical **age-hardening** heat-treatment, leading to precipitation of a fine dispersion of the Ti_2Cu-phase, as indicated in the phase diagram of Fig. 3.12. The strength is increased further if the alloy is cold worked prior to ageing. This, coupled with ready weldability makes it appropriate for use in, for example, gas turbine engine casing assemblies.

β-alloys

β-alloys require the addition of sufficient β-stabilising elements, such as vanadium, is indicated in the phase diagram of Fig. 3.13. The resulting b.c.c. structure is much more readily cold-formed than the hexagonal α-Ti. One example of this group is Ti–13V–11Cr–3Al, and final strengthening is achieved by age-hardening, which, together with solution

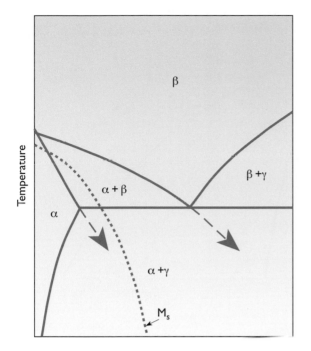

Fig. 3.12 Schematic phase diagram for age-hardening Ti alloys. M_S shows the martensite start temperature.

hardening by the β-stabilising elements can give tensile strengths in excess of 1300 MPa.

α/β-alloys

α/β-*alloys* have phase diagrams of the general form indicated in Fig. 3.13, and are of the greatest commercial importance, with Ti–6Al–4V comprising over half the world sales of Ti alloys. Their high strength, coupled with improved formability over the α alloys, have made them attractive contenders for forged components such as the fan blades of jet engines.

The microstructure and properties of the α/β alloys can be varied by heat-treatment. If the alloy is heated into the β-phase field and then quenched rapidly it undergoes a diffusionless phase transformation to form a martensitic phase, α'. This transformation starts at temperature M_s and finishes at temperature M_f, as indicated on Fig. 3.13.

The change of strength with heat-treatment is also indicated in Fig. 3.13. The annealed alloys gradually increase in strength as the alloying element content is increased: this (by the lever rule) will be accompanied by a progressive increase in the volume fraction of β-phase. Quenching to form α' produces a modest increase in strength (much less than that encountered in martensitic steels) – the highest value corresponding to that alloy whose M_f temperature is at room

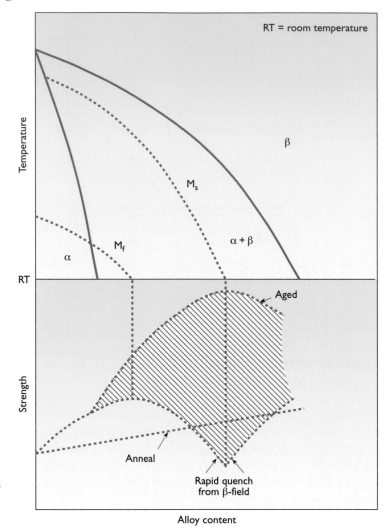

Fig. 3.13 Schematic diagram for the heat-treatment of titanium alloys (after P.H. Morton).

temperature, so the microstructure will be wholly transformed to α', with no metastable β persisting.

If the quenched alloys are then aged to decompose the retained metastable β, a moderate increase in strength is observed, as indicated in Fig. 3.13.

Table 3.5 lists typical mechanical properties of some titanium alloys.

Copper Alloys

Copper and its alloys are widely used because of their high electrical and thermal conductivity, corrosion resistance, and

Table 3.5 Typical mechanical properties of some titanium alloys

Alloy	Condition	Density Mg m^{-3}	Young's Modulus GPa	Proof stress MPa	UTS MPa	Elongation (% in 50 mm)
Commercially pure Ti (α) (IMI155)	Annealed	4.51	~105	540	640	24
Ti-6Al-4V (α/β) (IMI318)	Annealed	4.46	~106	925	990	14
Ti-13Mo-11Cr-3Al (β)	Aged	4.87	~106	1200	1280	8

ease of fabrication. Certain compositions for wrought products have counterparts among the cast alloys, so for a given component a designer can select an alloy before deciding on the manufacturing process. We will consider this family in **three** groups – the grades of copper itself, then the high-copper alloys and finally the alloys containing larger quantities of alloying elements.

Grades of Copper

Tough pitch copper contains residual oxygen from the refining process, including oxide particles, which makes it unsuitable for tube manufacture or for welding. *Deoxidised copper* is more appropriate for such applications, and small additions of phosphorus or another deoxidiser is made for this purpose. If residual deoxidiser remains in solid solution, the electric conductivity of the copper is impaired, so *high conductivity copper* is refined and deoxidised to a high degree of purity for use in electrical applications.

Copper Alloys of Low Solute Content

Arsenical copper contains up to 0.5% As, which has the effect of reducing the tendency of the metal to scale when heated, and also gives a slight increase in high-temperature strength through solute hardening.

Free-cutting copper contains ~0.5% Te which gives rise to the presence of second-phase particles of a telluride phase. During machining, these particles cause the swarf and chippings to break up into small fragments allowing the cutting fluid access to the interface between the workpiece and the tool. This increases tool life for a given rate of machining; sulphur-bearing alloys are also available for the same purpose.

Copper-beryllium alloys are age-hardenable, as suggested by the phase diagram of Fig. 3.14. Alloys contain typically 1.9% Be,

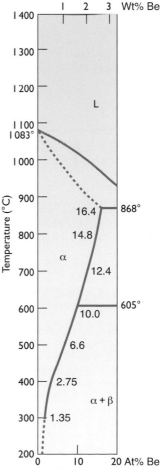

Fig. 3.14 The Cu–rich end of the Cu–Be phase diagram.

and are subjected to the normal sequence of solution treatment at 800°C, quenched and artificially aged in the range 300–320°C to precipitate the intermetallic phase CuBe. The alloys also normally contain <0.5% Co or Ni which segregates to the grain boundaries and inhibits the precipitation of coarse particles of CuBe in those regions. Tensile strengths of the order 1400 MPa can be achieved, but with elastic moduli only two-thirds that of steel. This allows large deflection for a given stress in such applications as springs, diaphragms and flexible bellows.

Copper Alloys of High Solute Content

Zinc, aluminium, tin and nickel are the most widely employed alloying elements in copper, and we will consider the commercial alloys in terms of these four groups.

The structure of the copper–zinc alloys, or **brasses** can be understood from the phase diagrams of Fig. 3.15. The zinc content can vary between about 5% and 40%; up to about 35% Zn single-phase α-solid solutions are formed with progressively increasing tensile strength due to solute hardening. The zinc also confers resistance to atmospheric and marine corrosion, as well as increasing the work hardening

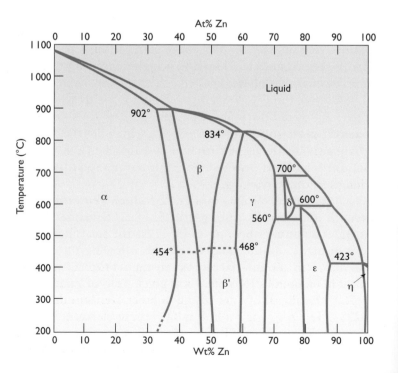

Fig. 3.15 The Cu–Zu phase diagram.

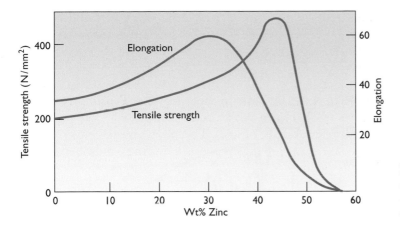

Fig. 3.16 Showing the change in tensile strength and ductility with Zn content of brasses.

rate. The α-brasses are thus particularly suited to cold working into wire, sheet and tube.

At higher Zn contents β-phase appears in the microstructure, which increases the tensile strength of the alloy but is associated with an appreciable loss in room temperature ductility. These effects are summarised in Fig. 3.16. At 800°C the β-phase is readily worked, however, the high Zn α/β brasses, such as the 60% Cu 40% Zn alloy ('60/40 brass', or Muntz metal) are well suited for shaping by extrusion and hot stamping.

The 'high tensile brasses' are formed from the 60/40 brasses by the addition of solution-hardening elements such as Al, Fe, Mn, Sn and Ni. These may be used for forgings as well as castings, notable among the latter being marine propellers. The single-phase α-'nickel-silvers' contain no silver, but 18–20% Ni has a decolorising effect on brass and confers good corrosion resistance. 'Nickel-brass' usually refers to the α/β alloy with about 45% Cu, 45% Zn and 10% Ni which is available as extruded sections.

Copper-nickel alloys form a continuous series of solid solutions, as shown in Fig. 1.11, so have essentially similar microstructures. The copper-rich alloys are used as condenser tubes – the higher the Ni-content the higher their mechanical strength and corrosion resistance. The addition of 1–2% Fe increases their resistance to impingement attack in moving sea-water.

Copper-tin alloys, or **bronzes** are again essentially α-solid solutions of Sn in Cu. The relevant phase diagram is shown in Fig. 3.17, and cast alloys may have tin contents in the range 5–19%. Tin is a strong solution-hardening element, but cast alloys are far from equilibrium and show cored microstructures and increasing volume fractions of the hard $(\alpha + \delta)$

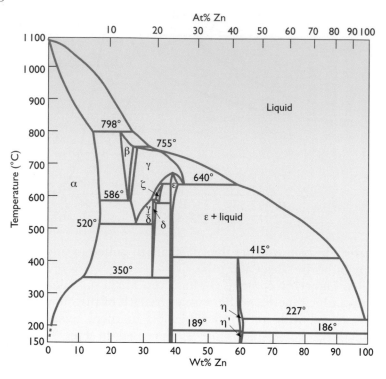

Fig. 3.17 The Cu–Su phase diagram.

eutectoid as the Sn content increases above about 7%. Typical applications include cast bearings and bushings, and **phosphor-bronzes** (containing 0.3–1% P) are employed when a bearing surface is required to bear heavy loads with a low coefficient of friction.

Copper-rich aluminium alloys are known as **aluminium bronzes**, the composition of the commercial alloys ranges from 5–11% Al, and the phase diagram of Fig. 3.18 enables the microstructures to be understood. The alloys are characterised by high strength coupled with a high resistance to corrosion and wear. Up to 7% Al the alloys consist of single-phase α, which is easily worked. They find widespread application in heat-exchanger tubing. The α/β alloys contain $\sim 10\%$ Al, and are either cast or hot-worked in the β-phase field. Slow cooling from β produces a eutectoid $(\alpha + \gamma_2)$ mixture, and quenching produces a hardened martensitic structure of β'. Although these changes are analogous to those found in steel (see later), such heat treatments are not widely applied to aluminium bronzes in practice.

The machining properties of brasses and bronzes are enhanced by the addition of a few % of lead which, being insoluble in both solid and liquid phases, appears as globules

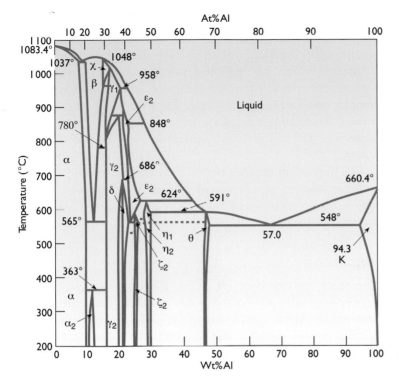

Fig. 3.18 The Cu–Al phase diagram.

Table 3.6: Typical mechanical properties of some copper alloys

Alloy	Condition	Density Mg m^{-3}	Young's Modulus GPa	Proof stress MPa	UTS MPa	Elongation (% in 50 mm)
Cu-1.85Be-0.25Co	Peak aged	8.25	159	1066	1205	7
Cu-30Zn α-brass	Annealed	8.55	100.6	115	324	67
Cu-40Zn α/β-brass	Hot rolled	8.40	102	130	371	40
Cu-31Ni-1Fe-1Mn	Hot rolled	8.94	152	114	369	50
Cu-10Sn-2Zn	Sand cast	~8.8	~111	139	286	18
Cu-8Al	Hot rolled	7.8	123	170	417	45

in the microstructure. These allow turnings to break up during machining.

Table 3.6 gives the properties of some copper alloys.

Lead Alloys

The most significant applications of lead and lead alloys are lead-acid storage batteries, and it is also widely used for building construction materials such as sheet because of its relative inertness to atmospheric attack. Lead is also finding increased application as a material for controlling sound and mechanical vibration, due to its high damping capacity.

Because of its low melting point (327.5°C), pure lead undergoes creep at room temperature, so it usually strengthened by small additions of solute. Thus 1.5 to 3% antimony or traces of calcium are often present in the plates of car batteries, and different manufacturers may use different solutes.

The important families of more concentrated alloys are the lead-tin solders (see later) and lead-based bearing alloys, which may contain antimony, tin and arsenic, for use in internal combustion engines.

Zinc Alloys

The main uses of zinc are as an alloying element (e.g. in copper which we have already discussed) and as a protective coating for steel (galvanising, etc.). However, zinc-based alloys are also used for the production of gravity castings, and pressure die-castings of high dimensional accuracy.

The casting alloys are based on the Zn–Al system (Fig. 3.19), and are close to the eutectic composition of 5% Al. The hypoeutectic alloys solidify with zinc-rich dendrites, whereas the hypereutectic contain Al-rich dendrites. The commercial alloys contain strengthening additions of copper and magnesium, but the greatest care has to be taken to prevent excessive pick-up of harmful impurity elements such

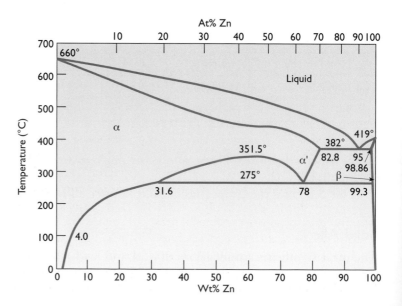

Fig. 3.19 The Al–Zu phase diagram.

Table 3.7 Typical mechanical properties of some lead, zinc and nickel alloys

Alloy	Condition	Density Mg m^{-3}	Young's modulus GPa	Yield stress MPa	UTS Mpa	Elongation (% in 50 mm)
Lead	> 99.99%	11.68	16.1	–	168	–
Zn-8.5Al-1Cu (ZA8)	Die cast	6.3	86	290	375	6–10
Ni-30Cu-1.5Fe-1Mn	Hot rolled	8.83	180	230	560	45
Ni-20Cr-2Ti-1.5Al Nimonic 80A	Fully aged	8.19	222	780	1220	30

as lead, cadmium, tin and iron. The alloys themselves are prepared from high-purity components, otherwise inter-granular embrittlement of the castings develops over a period of time.

The properties of lead, zinc and nickel alloys are presented in Table 3.7.

Nickel Alloys

Commercially pure nickel offers excellent corrosion resistance to reducing environments (in contrast to those metals and alloys which owe their resistance to the presence of a tenacious oxide film), and are found in the chemical processing industry, as well as in food processing applications. As an electroplated coating, nickel is widely used in the electronics industry.

Nickel–copper alloys (see Fig. 1.11) possess excellent corrosion resistance, notably in sea water. The Monel ($\sim 30\%$ Cu) series of alloys is used for turbine blading, valve parts and for marine propeller shafts, because of their high fatigue strength in seawater.

Nickel–chromium alloys form the basic alloys for jet engine development – the **Nimonic alloys**, or nickel-based superalloys. The earliest of these, Nimonic 80A was essentially 'Nichrome' (80/20 Ni/Cr) precipitation hardened by the γ' phase (Ni_3Ti,Al). Alloy development proceeded by modifying the composition in order to increase the volume fraction of the γ' phase; this made the alloys increasingly difficult to forge into turbine blades, the higher γ' volume fraction alloys had to be cast to shape. Figure 3.20 illustrates the historical development of these alloys in terms of their high-temperature strength, based on rupture in 1000 hr at 150 MPa. The limitation of the conventionally cast super-

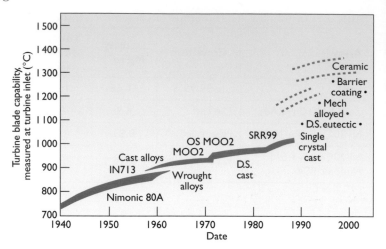

Fig. 3.20 Turbine inlet temperature versus year (after A. Kelly: *Proc. 10th Risø Internat. Symp. on Metallurgy and Materials Science*, 1989, 635).

alloys was their lack of creep ductility due to cavitation at the grain boundaries lying perpendicular to the maximum tensile stress, so *directionally solidified* (DS) and eventually *single crystal* alloys were developed.

Their maximum operating temperature is limited by the tendency of the γ' phase to return into solid solution, and so by choosing an insoluble phase, powder metallurgically produced *oxide dispersion-strengthened* (ODS) superalloys have been developed by a technique known as *mechanical alloying*. In spite of their high creep strength, shown in Fig. 3.20, the low ductility of present ceramics limits their use in this context.

Steels

We will approach our consideration of this large group of engineering alloys by dividing it into three sections: *low carbon steels, engineering steels* and *stainless steels*.

Low Carbon Steels

The iron-carbon phase diagram is shown in Fig. 3.21, and we may arbitrarily describe low carbon steels as those containing no more than 0.2% carbon. Their microstructure will thus be essentially ferrite (α) with a small volume fraction of carbide (Fe_3C) often present as the α/Fe_3C eutectoid (pearlite). The lower end of this range of carbon content (i.e. very low carbide content) is mainly applicable to steel in the form of thin strip and the upper end to structural steel in the form of thicker plates and sections.

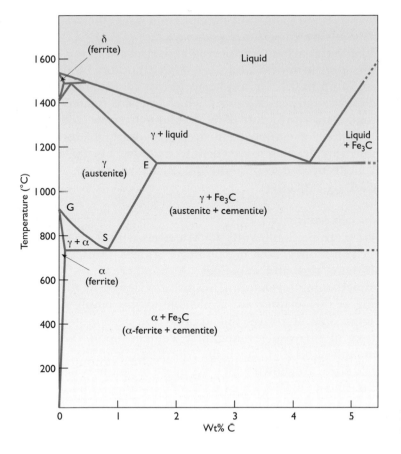

Fig. 3.21 The Fe–C phase diagram.

Strip Steels

These materials are manufactured by hot rolling to thicknesses not less than 2 mm, and then finally cold rolled to the required final thickness. Since cold forming (e.g. by deep drawing, stretch forming or bending) is a widely used application for low-carbon strip steels, a final recrystallisation anneal is applied. Considerable processing expertise is required to optimise the grain size, the crystallographic orientation ('texture') of the grains and the solute (C and N) content in order to enhance the formability of the strip. The presence of a strong yield point is undesirable, as the irregular flow may lead to a poor surface finish due to the presence of Lüders lines. Additions of niobium or titanium can be made to 'getter' the C and N solute atoms to form carbides or nitrides, thus improving the cold-forming response – the steels being known as *interstitial-free* (IF).

With mild steels of higher C content, small Nb, V or Ti additions lead to sufficient precipitation of finely divided precipitates during cooling from the hot-rolling temperature to give rise to significant precipitation strengthening. These materials are known as *High-strength Low-alloy* (HSLA) steels. Other approaches to achieving high strength in low-carbon strip steels include *solution strengthening*, notably by the addition of up to 0.1% phosphorus for use in automobile body pressings, and *dual-phase steels* which are heat-treated to form a mixed microstructure of ferrite and martensite (see p.103). The latter have a low yield strength but a high work-hardening rate, leading to excellent formability, but the alloying elements needed to promote this microstructure make them relatively expensive.

Structural Steels

These are predominantly C–Mn steels with ferrite-pearlite microstructures, used in buildings, bridges, ships, as well as off-shore rigs and pipelines. A number of the general strengthening mechanisms we have reviewed at the beginning of this chapter are applicable to this family of steels. Figure 3.22 illustrates this approach in accounting for the effect of increasing manganese content upon the tensile strength of 0.15%C steel: the contributions to the observed yield stress are shown which are made by solution-hardening of the ferrite by Mn, by grain-size hardening (equation 3.2), and by the increasing volume fraction of the hard eutectoid pearlite phase.

The properties of particular importance in these materials include *weldability* and *resistance to brittle fracture* under the service temperature conditions.

Engineering Steels

High strength steels might be classed as those with tensile strengths in excess of 750 MPa, and they embrace a wide range of composition and microstructure.

Heat-treated Steels

In many applications, the heat-treatment applied is that of **quenching** followed by **tempering**. The steel is heated into the γ-phase field, and then quenched in order to form the hard, metastable α'-phase known as **martensite.** The cooling-rate during quenching must be such that the entire cross-section of the specimen is converted to martensite : a TTT curve is illustrated in Fig. 3.23, where it may be seen that the cooling-rate at the centre of large cross-section may be insufficient to suppress the nucleation of other phases. Small quantities of Cr and Ni are added to **low-alloy steels**, which have the effect of moving the TTT curve to

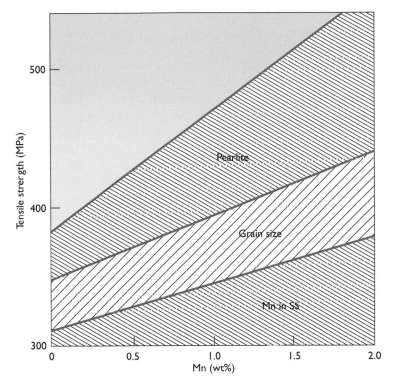

Fig. 3.22 Showing the effect of Mn, grain size and pearlite on the strength of a 0.15% C steel. (SS = solid solution).

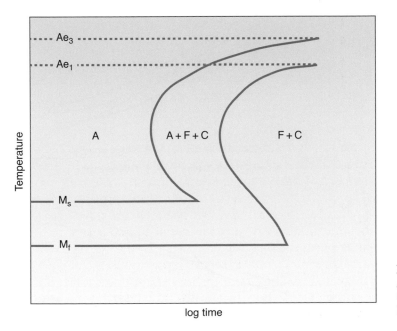

Fig. 3.23 A TTT curve. A = austenite, F = ferrite, C = carbide and M = martensite.

the right, thus enabling material of larger cross-section to be fully transformed to martensite. This is known as increasing the **hardenability**.

The martensite is finally tempered by heating in the range 200–600°C, causing it to decompose to a mixture of carbide particles in α, known as 'tempered martensite'. The fineness of the carbide dispersion depends upon the tempering temperature: the lower the temperature the finer the dispersion of carbide and the harder the material (Fig. 3.24). The diagram also demonstrates that as the strength increases the ductility decreases: this follows from the Considère construction (Fig. 2.5), since the strain for plastic instability (i.e. the point of contact of the tangent) decreases as the stress-strain curve is raised.

The progressive fall in hardness with increasing tempering temperature arises because of the rapidly increasing diffusivity of carbon, allowing progressively coarser particles of iron carbide to form in the steel. **High-speed steels** contain major additions of strong carbide-forming elements such as Cr, Mo, W and V. When these steels are quenched and tempered, alloy carbides form in the high temperature range when the hardness of plain carbon steels declines, so **secondary hardening** (Fig. 3.25) is observed. High speed steels are so-called because they maintain their hardness during high-

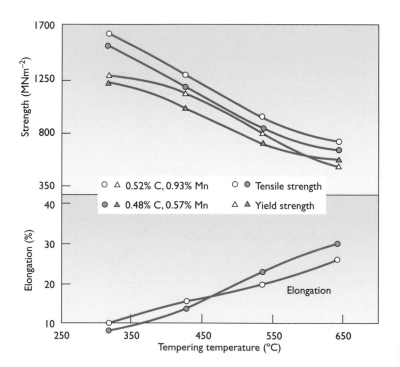

Fig. 3.24 The effect of tempering temperature upon tensile properties of a carbon steel.

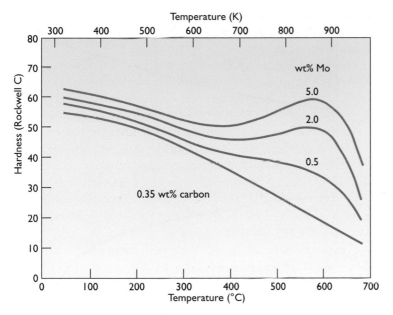

Temperature (K)

wt% Mo

5.0

2.0

0.5

0.35 wt% carbon

Hardness (Rockwell C)

Temperature (°C)

Fig. 3.25 Showing secondary hardening during tempering of a Mo-bearing steel.

speed machining operations: appreciable softening does not occur until the temperature exceeds about 550°C, which represents the maximum operating temperature for these steels.

Pearlitic Steels

As apparent from Fig. 3.21, air-cooled medium-to-high carbon steels undergo a eutectoid decomposition to form a microstructure of islands of lamellar pearlite ($\alpha + Fe_3C$) in an α matrix, and their strength will increase as the volume fraction of pearlite increases as illustrated in Fig. 3.26. Their applications include rail steels, as well as high carbon wire rod. The latter are transformed from γ at a temperature near the 'nose' of the TTT curve (Fig. 3.23) to form a microstructure of minimum interlamellar spacing. The wire is then cold drawn, and the work-hardened product may have tensile strengths in excess of 2 GPa.

Maraging Steels

These steels may be termed **ultra-high strength steels**, and may exhibit very high strengths with good fracture toughness but at a cost greatly exceeding that of conventional quenched and tempered low-alloy steels.

They generally contain about 18% nickel, but their carbon content does not exceed 0.03%. On air-cooling, they transform martensitically to a fine-grained α of high dislocation density. They are then age-hardened at around 500°C when a high density of intermetallic phase precipitates, due to the

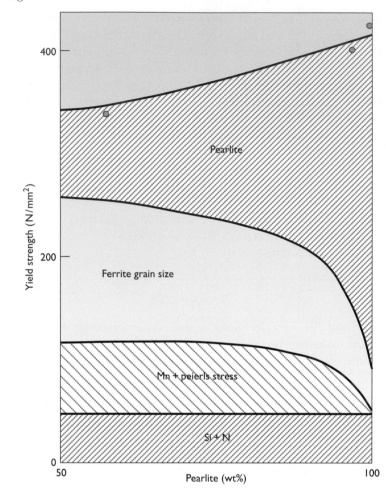

Fig. 3.26 Showing the contributions to the strength of pearlitic steels. Experimental points are shown •.

presence of Co, Mo and Ti in the alloy. They owe their high strength, therefore, to a combination of strengthening mechanisms.

Stainless Steels

With the addition of about 12% Cr, steels exhibit good resistance to atmospheric corrosion, and these *stainless steels* owe their passivity to the presence of a thin protective film of Cr_2O_3. They are used because of this corrosion and oxidation resistance, as well as their pleasing appearance. Because of the number of elements they contain, their microstructure cannot be simply represented on the simple phase diagrams we have considered, but they are normally classified according to their crystal structure as *ferritic, martensitic, austenitic,* and *duplex* stainless steels. The alloying elements present can be classified as ferrite stabilisers, which tend to promote the formation of the bcc phase, or as austenite stabilisers, which tend to promote the fcc γ-phase. The ferrite is normally referred to as δ-ferrite (Fig. 3.21) in these steels. In predicting the room temperature microstructure of stainless steels, therefore, the balance between the ferrite and austenite formers has to be considered. An empirical and approximate approach to this question can be made by means of the **Schaeffler** diagram, which has been modified by H. Schneider (*Foundry Trade Journal* 1960 **108**, 562) illustrated in Fig. 3.27. This indicates the structures produced after rapid cooling from 1050°C, and the axes are the *chromium equivalent* and the *nickel equivalent*. The former indicates the proportion of elements (expressed as weight percentage) that behave like chromium in promoting ferrite according to:

Cr equivalent =
$$Cr + 2Si + 1.5Mo + 5V + 5.5Al + 1.75Nb + 1.5Ti + 0.75W$$

Similarly, the austenite formers give:

Ni equivalent =
$$Ni + Co + 0.5Mn + 0.3Cu + 25N + 30C$$

The three most important groups are as follows:

Ferritic stainless steels are generally of lower cost than the austenitic steels, and are used when good cold-formability is required. About half of the stainless steels produced are rolled to sheet which is subsequently cold-drawn into articles such as cooking utensils, sinks, automotive trim, etc.

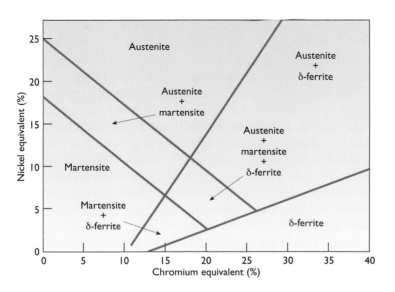

Fig. 3.27 Schaeffler diagram showing the structures of various stainless steels.

Martensitic stainless steels contain more C than the ferritic, and can be heated to form austenite then cooled with excellent hardenability to form martensite. They can then be tempered to yield strengths in the range 550 to 1860 MPa and can be used for cutting implements.

Austenitic stainless steels are non-magnetic, because of their fcc structure, and show no ductile/brittle fracture transition. Annealed type 304 stainless steel has a yield strength of about 140 MPa and a UTS of about 585 MPa, so work-hardening is an obvious means of strengthening. Solid solution strengthening can be employed in these steels, and N is most effective in this regard – the 200 series of Cr–Mn–Ni–N steels being an example.

The steels are readily welded, but care must be taken to prevent the precipitation of chromium carbide in the grain boundaries. This may occur in the 'heat-affected zone' adjacent to the weld, leaving the material adjacent to the grain boundary depleted in chromium and thus no longer corrosion-resistant. Intergranular corrosion may thus occur is these regions – a phenomenon known as 'weld decay'. The problem may be overcome by employing steel of very low (< 0.06%) carbon content, so that chromium carbide cannot form. An alternative solution is to employ a 'stabilised' stainless steel containing an element with a higher affinity for C than Cr, such as Ti or Nb. In the heat-affected zone of the weld preferential precipitation of NbC or TiC rather than chromium carbide occurs in the grain boundaries, and the grain boundaries are not depleted in Cr and remain impervious to attack.

Cast Irons

As their name implies, cast irons are shaped by casting into a mould rather than by forging in the solid state. They are alloys of iron that usually contain between 2.5 and 4% carbon (and from 1 to 3% silicon, which tends to promote the appearance of the carbon as *graphite*, rather than as carbide, Fe_3C), and from the phase diagram of Fig. 3.21 it is clear that a binary Fe–4% C alloy will be close to the eutectic composition and thus have a low melting temperature. Cast irons are very fluid when molten and have good casting characteristics, but the castings usually have relatively low impact resistance and ductility, which may limit their use.

The mechanical properties of cast irons depend strongly on their microstructure, and there are three basic types – white iron, grey iron, and ductile iron.

White Cast Iron

This has a low silicon content, and may contain carbide-stabilising elements (such as Cr). When cooled rapidly from the melt, graphite formation does not occur, and the microstructure may be predicted from Fig. 3.21 to consist of dendrites of austenite in a matrix of eutectic (iron carbide + austenite). On cooling to room temperature, the austenite decomposes to eutectoid pearlite.

The absence of graphite in the microstructure results in a completely white fracture surface – hence the name of the material. The hardness is high (400–600 H_v) due to the presence of the carbide and the pearlite, and this makes white cast iron very suitable for abrasion-resistant castings.

Grey Cast Iron

In this iron, the eutectic which forms consists of flakes of *graphite* + austenite. The formation of graphite rather than iron carbide is promoted by the presence of silicon and by conditions of slow cooling. If the casting consists of varying sections, then the thin regions will be 'chilled' and cool at a greater rate than the thick regions, and only the latter will form grey cast iron.

The iron-graphite phase diagram is similar in form to Fig. 3.21, with C (graphite) replacing Fe_3C, and it may still be used to account for the microstructures that develop. Hypo-eutectic irons solidify to dendrites of austenite in an austenite/graphite eutectic, and hyper-eutectic irons form primary graphite flakes in the same austenite/graphite eutectic. On further cooling, the austenite decomposes at the eutectoid temperature – at high cooling rates to pearlite and slow cooling rates to a ferrite-graphite eutectoid.

The microstructure of a grey cast iron is illustrated in Fig. 3.28. These irons are so called because the presence of graphite in the microstructure leads to a grey-coloured fracture surface.

Grey cast iron is an attractive engineering material because of its cheapness and ease of machining. Graphite is devoid of strength, but its formation tends to compensate the tendency for castings to shrink on solidification. Its flake-like form effectively means that the microstructure is full of cracks, with the result that the material exhibits little ductility, although it is strong in compression. Graphitic irons exhibit a high *damping capacity*, since vibrational energy is dissipated at these internal interfaces.

The **carbon equivalent value** (CE) is an index which combines the effect of Si and P upon the eutectic of iron and

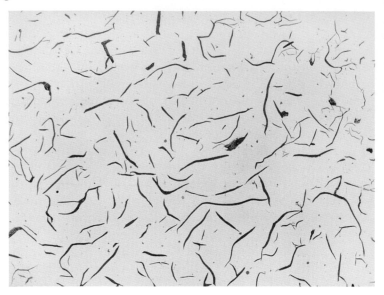

Fig. 3.28 Optical micrograph of grey cast iron (unetched) × 500.

carbon, and is defined:

$$CE = Total \%C + (\%Si + \%P)/3$$

and it determines how close a given composition of iron is to the theoretical iron-carbon eutectic (CE = 4.3%), and therefore how much free graphite is likely to be present for a given cooling rate. The mechanical properties of a cast iron are largely controlled by the graphite content, and Fig. 3.29 indicates the relationship between CE value, UTS and section thickness (i.e. cooling rate) and the resultant microstructure. It is seen that white cast iron (hard and unmachinable) is formed at low CE values and rapid cooling, and grey irons (first pearlitic and then ferritic) as CE values and section sizes increase.

Ductile Iron

Malleable Cast Iron may be produced by taking a white iron casting and holding it at 950–1000°C, when breakdown of the iron carbide occurs:

$$Fe_3C \rightarrow 3Fe + C$$

producing graphite in the form of aggregates known as 'temper carbon'. These are not in flake form, and are thus much less deleterious to the mechanical properties and the result is a material which possesses a good measure of strength combined with ductility.

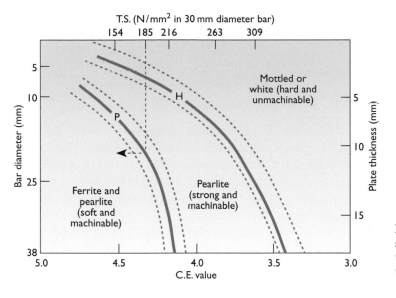

Fig. 3.29 Showing the structures of cast irons of various Carbon Equivalent value and section thicknesses.

Nodular, or Spheroidal Graphite (SG) Cast Iron contains graphite spheroids in the as-cast state, by the addition of cerium and/or magnesium to the iron, as illustrated in Fig. 3.30. SG irons can be made with a pearlite matrix or ferrite with appropriate heat-treatment, or with acicular or austenitic matrix when suitably alloyed, and they behave as more or less normal ductile ferrous materials.

In view of the large number of compositions of steels and cast irons commercially available, and the fact that their properties may be varied over a wide range by appropriate

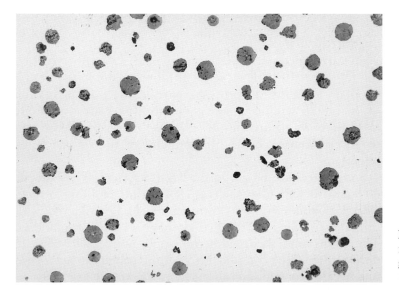

Fig. 3.30 Optical micrograph of spheroidal graphite cast iron (unetched) × 500.

Table 3.8 Density and Young's Modulus of ferrous materials.

Alloy	Density (mg m^{-3})	Young's modulus (GPa)
Structural steels	~7.9	~210
Grey cast irons	~7.05	100–145

heat-treatment, it is impossible to present a brief summary table of the properties of these materials as we have attempted to do in the case of non-ferrous alloys. In the Appendix we give an outline of the sources of material property data, and reader must be guided by this when seeking information on the properties of ferrous alloys. Table 3.8 simply gives an indication of their range of density and Young's modulus.

JOINING OF METALS AND ALLOYS

Metals may be joined together by *mechanical fastening*, by *soldering or brazing*, which employs a layer of metal of lower melting-point than the metals to be joined, by *welding,* in which the basis metals are fused, and by *adhesive bonding,* which employed a non-metallic layer.

Mechanical Fastening

The important process of rivetting will not be considered here. One important consideration regarding such joints is the question of whether corrosion problems arise in their vicinity, and this aspect will be referred to below.

Soldering and Brazing

These are processes whereby the basis metals are wetted by the filler metals, with subsequent filling of the joint gaps by capillary action. If the process is carried out in air, oxidation of the metals will occur and therefore fluxes are commonly used in both techniques to dissolve the oxide films and to ensure wetting by the filler metal of the metals to be joined.

Soft solders are based on the Sn–Pb system. The eutectic 60% Sn- 40% Pb alloy melts at 183°C, thus enabling joints in electrical and electronic components to be soldered with minimum danger of thermal damage to the components. The joints may be readily reheated where necessary to disassemble the components and resoldered, as when making repairs and alterations to an electronic assembly.

The shear strength of a soft-soldered lap joint depends on the local state of stress, but is usually in the range 20–40 MPa. Alloys of lower Sn content than the eutectic are used where ever it is possible to use a higher soldering temperature, with 50% Sn used in copper plumbing, and 40% Sn in lead plumbing, where the longer freezing range allows the solder to pass through a 'mushy' stage as it cools, enabling smoothing of the joint by 'wiping'. Solders with tin contents between 30% and 5% (with Sb and As additions) are used as filler metal in repairing or shaping automobile bodywork.

If higher strengths are required, then *brazed* joints are employed.

Al-Based Brazing Alloys
Alloys based on the Al–Si system are used for brazing Al and certain Al alloys, using chloride-fluoride mixtures as a flux in order to remove the tenacious oxide film present with these materials.

Cu-Based Brazing Alloys
These are widely used for joining both ferrous and nonferrous basis metals, and fall into the three classes of virtually pure copper, copper–phosphorus (which is self-fluxing in many cases) and copper-zinc (brass), which often employ a boric acid or borate flux.

Ni-Based Brazing Alloys
These embrace a wide range of compositions, which may include Cr, P and Si, and they find their widest use in assemblies of stainless steels and of nickel or cobalt alloys. Many of the brazed joints retain good strength at temperatures approaching 1000°C.

Brazed joints of all composition are significantly stronger than those in soft solder, with copper-based joint strengths being in the range 250–400 MPa and nickel-based joints in the range 300–600 MPa, thus often equalling that of the metals being joined. It is however to welded joints that the highest strengths are normally encountered.

Welding

An ideal weld would be chemically and physically indistinguishable from the bulk material: this may sometimes be approached in **solid state welding**, but seldom in **fusion welding**.

Solid State Welding

The formation of a sound joint requires either chemically clean surfaces or sufficient deformation to squeeze out any contaminants such as surface oxides. Cold pressure welding works well with ductile metals such as aluminium and copper, provided the surfaces to be joined are carefully prepared to be free from contamination. In hot pressure welding the material adjacent to the weld is softened, the two surfaces are forced together, thus squeezing out any surface contaminants. The heating may be externally applied, or, in friction welding, the heat is generated by rotating one surface against the other. The latter technique is widely used in the automotive industry for the manufacture of welded drive shafts.

'Diffusion bonding' can be employed to bond two surfaces without recourse to plastic deformation, but it is a relatively slow and expensive process which is only used in special applications.

Fusion Welding

'Welding Handbooks' are readily available to the engineer to provide a guide to appropriate welding processes for given alloy compositions, joint design and joint size. Here we will discuss some of the factors that may affect the microstructure and properties of fusion welds.

The Weld Metal

The weld metal is essentially small casting, and the essentials of its structure can be appreciated by referring to our earlier discussion of the mechanism of crystallisation of metals and alloys in Chapter 1. Cored columnar crystals form on the still solid component surface along the fusion line. They then grow along the direction of steepest temperature gradient in the weld pool, as indicated in Fig. 3.31. The crystals which grow from the melt initially share the same orientation as the solid, so it is important to consider whether grain growth occurs in the component material adjacent to the fusion line, as this will influence the grain size of the solidified weld metal. In fact, the grain size in the weld is controlled by the grain size at the fusion line, since the grain boundaries there will be common to both. The region of the component adjacent to the weld is known as the heat-affected zone (HAZ).

The Heat-Affected Zone

The temperature-distance profile across the weld is a result of the balance between the rate of heating and the rate at which

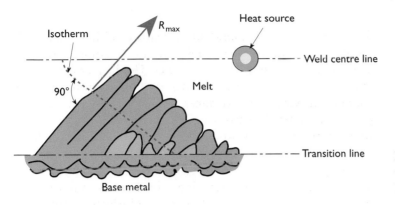

Fig. 3.31 Showing growth of columnar crystals in the weld pool.

heat is conducted away – the base material forming a large heat sink. A more intense heat source will give a steeper profile and the HAZ will be confined to a narrower region. The changes in microstructure that take place in the HAZ will depend on the material being welded, and upon its thermal and mechanical history. For example, if the material is in the cold-worked condition it will recrystallise, and if it is age-hardened the precipitate distribution will be changed. Different changes will occur in different temperature regimes of the HAZ, and Fig. 3.32 is a schematic diagram illustrating the possible regimes in the HAZ adjacent to a weld in a low-carbon steel. The adjacent iron-carbon phase diagram indicates the temperatures concerned in the case of a 0.15 wt% C steel, and the various regimes are self-explanatory.

Residual Stresses and Cracking
When a weld is deposited, the volume of material being heated and cooled is small relative to the whole assembly, so that as the weld and HAZ cool and contract they are constrained by the surrounding unaffected material. Large stresses develop, leading to local plastic deformation, and when all the structure returns to room temperature, tensile residual stresses remain in the weld. The magnitude of these stresses can be reduced by preheating the structure, or the stresses relieved by a post-weld heat-treatment – neither of which process is straightforward in the case of large structures.

These internal stresses can lead to cracking in the weld deposit and HAZ through several mechanisms:

(a) Solidification cracking.
This occurs in the weld deposit during cooling, and is found at the weld centre line or between columnar grains.

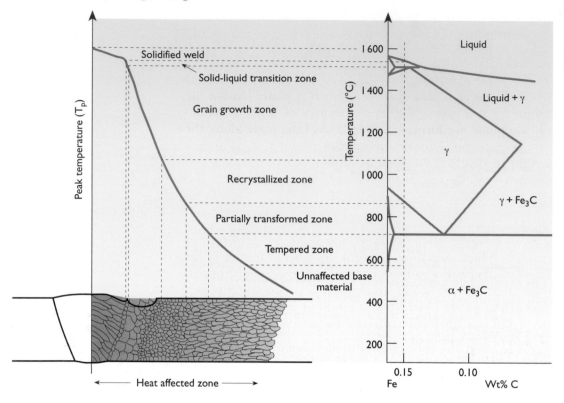

Fig. 3.32 Schematic diagram of the heat-affected zone of a 0.15 wt% C steel indicated on the Fe–Fe₃C phase diagram.

(b) Hydrogen-induced cracking in steel.

Atomic hydrogen can be introduced into the weld during the welding process – its principal origin being moisture in the electrode fluxes employed, although hydrocarbons on the plate being welded is another possible source. Hydrogen can give rise to so-called *cold cracking* in the HAZ (underbead; root crack) or in the weld metal itself, and is the most serious and least understood of all weld-cracking problems.

The solubility of hydrogen in austenite is much higher than in ferrite or martensite, so that if a steel transforms from austenite on cooling, it will be highly supersaturated with respect to hydrogen. Under these conditions, hydrogen diffuses to discontinuities in the metal such as grain boundaries and nonmetallic inclusions where it recombines to form hydrogen gas as microscopic bubbles which can develop into cracks.

(c) Liquation cracking.

This occurs in the HAZ near the fusion line, and is associated with the segregation of impurities such as sulphur and phosphorus to melted grain boundaries during welding. On cooling, these segregants tend to form films of intermetallic compounds, and with the development of high residual stresses these impurity-weakened boundaries tend to rupture.

(d) Lamellar tearing.

This occurs just outside the HAZ, and is commonly observed when a weld runs *parallel to the surface* of a plate. During the rolling of steel plate, flattened stringers of MnS or oxide-silicate phase are formed in the plane of the plate along the rolling direction. The orientation of the residual stress in such a weld is such that the weak inclusion/matrix interface can decohere and thus nucleate a crack.

Adhesive Bonding

This is brought about by applying adhesive to prepared surfaces which are then brought together. Heat may be applied to encourage adhesive setting, although the temperatures employed are usually too low to affect the structure and strength of metal parts. This method of joining provides a more uniform distribution of stress and a larger stress-bearing area than conventional mechanical fasteners such as rivets or bolts. The adhesive must wet the substrate to ensure a good bond, and its presence minimises or prevents electrochemical corrosion between dissimilar metals (see later). There is also evidence that adhesive bonded joints show an improved resistance to fatigue loading, compared to welded joints.

The upper limit of service temperature of such joints is usually 175°C, although materials are available which permit limited use up to 370°C.

Surface Preparation

Chemical surface preparations which enhance adhesive bonding have been available for many years. In the case of aluminium and titanium, these preparations not only clean the surfaces but also produce porous oxide structures with distinctive morphologies which allow the adhesive to penetrate into the metal/oxide interface. This 'mechanical interlocking' of hardened polymer with the substrate increases the bond strength and durability.

Iron does not form coherent adhesive oxides, so that steel is usually only chemically cleaned. The application of a surface coating such as zinc phosphate or metallic zinc ('galvanising') has been found to enhance the properties of an adhesive bond.

Types of Adhesives

Epoxy and acrylic adhesives are the most widely used structural adhesives. These are thermosetting polymers, but for many steel applications the size of the parts preclude the prolonged heating and high pressure cure cycles that are required by high performance epoxies.

In the automotive industry, for example, adhesive bonding has been employed principally in 'non-structural' , i.e. non-load bearing applications. These include the bonding of friction linings to braking components with phenolic-based adhesives because of the high temperature requirement. Polyurethanes are employed to bond the fixed glazing to the body shell, which confers a significant increase in body stiffness compared to the use of rubber gaskets. Adhesives are now also widely used in the bonding of the inner to the outer panels of doors.

Most of the adhesives intended for bonding metal body-shell assemblies require heat curing, and this is most effectively carried out during the paint curing cycle. *Reactive hot melt* adhesives have been developed which are applied at elevated temperature and, on cooling, solidify to give some initial strength in order to survive the early stages of the paint process. These adhesives, based on epoxy, polyurethane or synthetic rubber (eg. polybutadiene), acquire their full strength during the high temperature paint-curing cycle.

DEGRADATION OF METALS AND ALLOYS

Metallic engineering components may degrade by chemical attack from their environment, that is by *oxidation* and *corrosion*, or by physical attrition due to *wear*. We will consider these situations in turn.

Oxidation of Metals and Alloys

Oxidation is important when alloys are used at elevated temperatures: a surface scale forms on the material and continued oxidation proceeds by the diffusion of metal ions, possibly oxygen ions and certainly electrons through the oxide layer. The oxidation process for the formation of a coherent film (i.e. one free from cracks and cavities) can be represented schematically by a primary cell as shown in Fig. 3.33.

The EMF (E) is related to the free energy of formation of the oxide (ΔG) by the equation:

$$E = -\Delta G / NF \qquad (3.8)$$

where N is the number of moles and F is Faraday's constant.

If the resistance of the cell to ionic migration is R_i and to electronic migration is R_e, then the rate of film thickening will

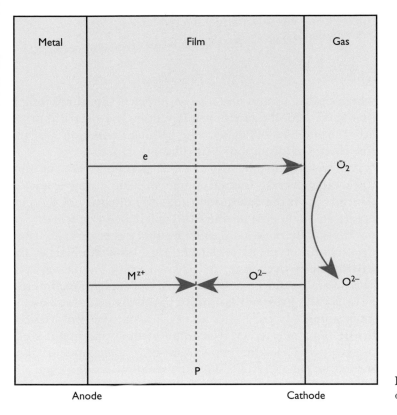

Fig. 3.33 Schematic oxidation process.

depend on the current flowing (I), given by Ohm's Law:

$$I = E/(R_i + R_e) \qquad (3.9)$$

The movement of electrons, cations and anions all contribute to the current in proportions represented by their respective transport numbers (n), where

$$n_a + n_c + n_e = 1 \qquad (3.10)$$

If the electrical conductivity of the film material is κ, for a film of area A and thickness y we can write:

$R_e = y/\kappa n_e A$; $R_i = y/\kappa$, $(n_a + n_c) \, A$, and equation 3.9 may be written

$$I = E\kappa A \, n_e \, (n_a + n_c)/y \qquad (3.11)$$

If $\mathcal{J}$ is the equivalent weight of the film substance, whose density is ρ, then a current I in time dt produces a volume of film given by:

$$I \, \text{dt.} \, \mathcal{J}/F.\rho$$

which is equal to $A \, dy$.

Substituting equation 3.11 for I, we obtain:

$$\mathrm{d}y/\mathrm{d}t = E\kappa A\, n_e\, (n_a + n_c)\, \mathcal{J}/\, F.\rho.y \qquad (3.12)$$

Thus
$$y^2 = kt$$

where k is a constant, and a *parabolic law* of film thickening is predicted, with the rate dependent upon the electrical properties of the film substance and (through equation 3.8) the free energy of formation of the film substance.

Almost any metal will obey a parabolic law of film thickening over a limited range of temperature, and it depends upon the assumption that the integrity of the film is perfect. In the first instance this will depend on whether or not the volume of the oxidation product is greater or less than the volume of metal replaced. This ratio, known as the **Pilling-Bedworth** ratio is less than unity for metals such as K, Na, Ca, and Ba, whereas metals of engineering interest have a ratio greater than unity, implying that most oxide films commonly encountered are in a state of compression. Brittle films may crack from time to time under this compressive stress, leading to 'quasi-linear' behaviour as illustrated in Fig. 3.34. Rectilinear oxidation occurs when the film is non-protective, and *logarithmic* behaviour occurs when the films develops flaws parallel to the interface, which limit

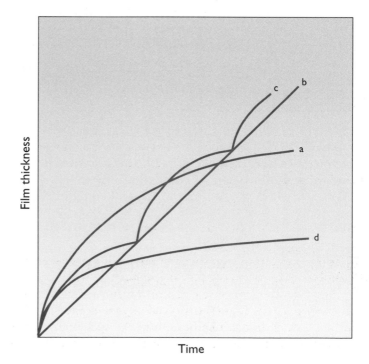

Fig. 3.34 Growth laws for oxide films on metals: (a) parabolic, (b) rectilinear, (c) quasi-rectilinear and (d) logarithmic.

the effective area available for diffusion. Logarithmic laws are also observed at low temperatures, possibly due to the reduced electrical conductivity under these conditions.

Improvement in oxidation resistance may therefore be expected if the driving force, $E = -\Delta G/nF$, is reduced, by making the alloy more 'noble'. The addition of 40% Cu to Ni to make the alloy Monel increases the oxidation resistance, as does the addition of Ni to Fe (as in 'Ni-resist' cast iron).

The second approach is to reduce the electronic and ionic conductivities of the film itself, as well as its resistance to cracking and spalling. The inherently protective film on Ti, Zr and Ta arises for this reason, but it must be borne in mind that these metals will not exhibit inertness under *reducing* conditions which may remove the protective film. Al-based alloys usually have an inherently protective film, although the addition of Cu to the alloy usually leads to a fall in its oxidation resistance.

Corrosion of Metals and Alloys in Aqueous Environments

Although we will use examples of metals being *immersed* in a liquid, the corrosion processes which occur are also encountered under conditions of ordinary atmospheric exposure, when the relative humidity is high. Condensation of moisture upon the metal surface then takes place, and degradation by 'wet corrosion' is observed.

It is well known that the mechanism of aqueous corrosion is electrochemical: a local **anode** is formed where the metal enters solution:

$$Me \rightarrow Me^{2+} + 2e \qquad (3.13)$$

Electrons flow through the metal and are neutralised at a local **cathode**. There is a variety of possible reactions for this electron discharge, and one of the most important involves the absorption of oxygen, which can be written as:

$$O_2 + 2H_2O + 4e \rightarrow OH^- \qquad (3.14)$$

So OH^- ions are produced in the liquid at the cathode, which thus becomes locally alkaline.

Let us consider an iron specimen partially immersed in an electrolyte (Fig. 3.35). Atmospheric oxygen will dissolve in the liquid, so the iron just beneath the surface of the liquid will experience the highest dissolved oxygen content and will thus act as a cathode. The bottom of the specimen, furthest removed from the surface, will experience least dissolved

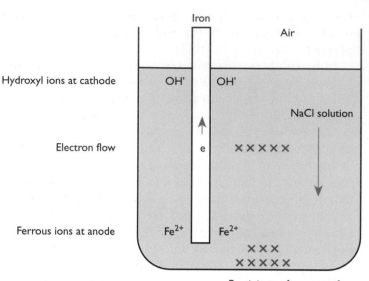

Iron

Air

Hydroxyl ions at cathode   OH'   OH'

NaCl solution

Electron flow   e   × × × × ×

Ferrous ions at anode Fe^{2+} Fe^{2+}

× × ×
× × × × ×

Fig. 3.35 A simple corrosion cell.

Precipitate of rust away from
both anode and cathode

oxygen and will act as an anode. Fe^{2+} ions willl thus pass into solution (i.e. the metal will corrode) in this region, equation (3.13), and electrons will pass up the specimen and be discharged near the surface, where OH^- ions will form in the electrolyte, by equation (3.14).

Diffusion will occur in the liquid over a period of time, and the Fe^{2+} ions and the OH^- ions will interact to form an iron hydroxide ('rust'). Two points are significant here: firstly there is the geometrical fact that the metal goes into solution in one region, electrons are discharged in another region – that where oxygen is most readily available – and the corrosion product is formed in a third place. Since the corrosion product does **not** form at the site where the metal is dissolving, there can be **no stifling** of the attack (as in the case of dry corrosion, when a protective film progressively reduces the rate of oxidation).

The second point to emphasise is the importance of the gradient in oxygen concentration in the electrolyte: this phenomenon of **differential aeration** is the origin of the EMF of the corrosion 'cell', and is in practice a very common source of corrosion in iron and steel. This is the reason why corrosion is often concentrated in crevices in structures – the regions of minimum oxygen availability forming anodes. Conversely, if oxygen is totally excluded from the system, electrons cannot be discharged by the process of equation 3.14, and so corrosion ceases. This is why steel wrecks immersed in very deep sea-water (e.g. the *Titanic*) survive for long periods, whereas those on or near the seashore quickly disintegrate by corrosion since they are so well aerated.

If two different metals are in electrical contact in an electrolyte, a **galvanic** EMF may exist and corrosion of the more reactive metal can proceed preferentially, even in the absence of oxygen. An example of this would be a galvanic couple between Al and Fe in an electrolyte, where the Al would form the anode and be preferentially dissolved. The *relative areas* of anode and cathode in contact with the electrolyte are important here – most dangerous being a situation where area of the anode is much smaller than that of the cathode: this leads to a high local intensity of anodic attack. For example, it would be unwise to fasten a steel plate with aluminium bolts or rivets since, under corrosive conditions rapid attack of the bolts will occur.

Galvanic corrosion can be employed deliberately to reduce the corrosion of metals such as iron by coupling them to more reactive ones (such as Zn or Al) which corrode 'sacrificially'. In terms of their susceptibility to galvanic attack in seawater, metals and alloys have been empirically grouped as follows:

Galvanic series for alloys immersed in seawater

Titanium alloys
Nickel alloys
Stainless steels
Silver alloys

Copper alloys

Lead and tin alloys

Cast irons
Structural steels
Cadmium

Zinc alloys

Aluminium alloys

Magnesium alloys

Galvanic effects are negligible between alloys in the same groups, but become increasingly more pronounced with alloys that are widely separated in the table.

Galvanic corrosion may be remedied by preventing electrical contact between the two metals by means of bushes or washers, for example. Controlling corrosion arising from differential aeration may be more complicated, and there are various means of effecting this.

Soluble inhibitors. In closed systems (e.g. when the corroding liquid is recirculated), the liquid may be treated with *soluble inhibitors*, which are of two main categories: the first is a reagent which removes oxygen from the solution, and second is one which leads to the formation of a *passive film* on the surface of the metal, thus stifling attack.

Cathodic protection. An example of this process is 'sacrificial protection' referred to earlier, whereby the metal to be protected in connected electrically to a more reactive metal in the galvanic series. *Galvanising* steel with a layer of zinc works in this way, as do slabs of Zn, Al or Mg that are attached at intervals to buried steel pipelines or to marine structures. An alternative method of obtaining cathodic protection is to use an *impressed current* from a suitable DC source: the steel to be protected is connected to the negative terminal and a inert metal anode is placed nearby.

Paints and lacquers. The main aim of these surface coatings is to exclude water and air from the metal surface. In many cases the exclusion is not total, and a paint layer may be regarded as introducing a large ionic resistance into the corrosion 'cell', thus reducing the corrosion current and hence the rate of attack. Some pigments used (such as red lead, Pb_3O_4) may act as inhibitors, while others (such as primers containing metallic zinc powder) is essentially a 'sacrificial' pigment.

Stress, Strain and Corrosion

Both 'dry' and 'wet' corrosion may be exacerbated in the presence of applied stress or strain.

Fretting corrosion. This occurs by two surfaces rubbing together: any surface oxide film will tend to be abraded away exposing the metal beneath to further environmental attack. Severe wastage to metal can occur in this way – the debris oxidising to form an abrasive powder. The situation of a loose joint under varying load may be susceptible to fretting corrosion, and also the case when wire ropes rub together in service over a protracted period.

Stress-corrosion cracking (SCC) may appear as mechanical failure by cracking under circumstances where, in the absence of corrosion, no failure would have been expected.

The stress may arise by external application or be present as a residual internal stress, and the corrosive medium is highly specific to each alloy. Under fluctuating stress, the phenomenon is known as *corrosion-fatigue*. Some form of surface pit must initially be formed, but there appears to be no common mechanism of crack propagation for all the different systems of alloys and corrodents that produce cracking. The rate of propagation of a stress-corrosion crack (da/dt) as a function of stress intensity (K) varies in the manner shown in the idealised curve of Fig. 2.20.

At low K values the crack velocity is highly K-dependent, becoming vanishingly small at $K_{I\,SCC}$. Considerable experimental patience is required to establish the latter value, which may correspond to crack growth rates of 10^{-11} m s^{-1}! There is a K-independent velocity plateau at intermediate values of K, and stage III in Fig. 2.20 shows a rapid acceleration as the value of K_{Ic} is approached. It is striking that an aggressive environment can commonly reduce the value of K_{Ic} by 95% to that of $K_{I\,SCC}$. The problem is normally tackled by metallurgical means, i.e. by modifying the composition of the alloy to increase its corrosion resistance, rather than by changing the state of stress or the environment.

Hydrogen embrittlement is a special case of SCC due to absorbed hydrogen, and is particularly important in the cases of high-strength steels and aluminium alloys. There are numerous potential sources of hydrogen, including welding, electroplating and corrosive attack.

In the case of ferritic steels, it has been suggested that dissolved hydrogen atoms can diffuse to internal cavities and recombine to form hydrogen gas. The gas can exert an internal pressure to assist the fracture process. An alternative suggestion is that the H atoms dissolve preferentially in regions of high triaxial stress (such as exist at the tip of cracks), where they locally lower the interatomic cohesion of the lattice.

There does not appear to be a single origin of hydrogen cracking – different micromechanisms of failure have been proposed for various materials, and further research is required before a full understanding is achieved.

Wear of Metals and Alloys

Wear occurs when two solid surfaces slide over each other, either with or without a lubricant, and empirical sliding wear tests have been devised to simulate practical situations and to

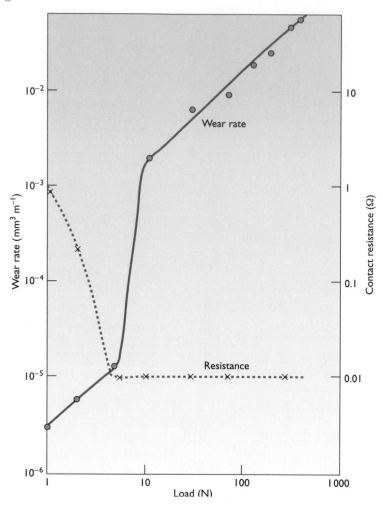

Fig. 3.36 Wear rate and electrical contact resistance for a leaded α/β brass pin sliding against a hard stellite ring as a function of normal load. (After Hirst and Lancaster).

provide design data on wear rates. A common test rig employs a pin pressed against a rotating disc or ring, the pin being the specimen and the rotating surface the *counter-face*. Several test rig methods are the subject of national standards, and contact stresses, thermal conditions, sliding speeds and chemical environment are all significant parameters in any wear test.

Wear is assessed by measuring the change in specimen dimensions during the course of the test, expressing the data as the amount of materials removed per distance slid as a function of the normal load. Figure 3.36, due to Hirst and Lancaster, shows on logarithmic axes the wear rate of a leaded α/β brass pin sliding against a hard stellite ring as a function of load, and the pattern of behaviour typifies that in many metallic systems in both lubricated and unlubricated conditions. The electrical contact resistance between the pin and the ring is also plotted, which allows the extent of metallic contact to be estimated.

There are seen to be two regimes of wear: *mild wear* at low loads and a transition to *severe wear* at higher loads. In each regime the data can be expressed in terms of the **Archard wear equation**:

$$Q = KW/H \qquad (3.15)$$

where Q is the wear rate per unit sliding distance, W the normal load, H the indentation hardness of the softer surface, and K is a dimensionless constant known as the *wear coefficient*. The quantity K/H is given the symbol k and called the *dimensional wear coefficient*, and it represents the volume of material removed by wear per unit distance slid.

In Fig. 3.36, the region of mild wear corresponds to the sliding surfaces being separated by oxide films, with only occasional direct metallic contact – hence the relatively high contact resistance. The debris formed consists of mixed oxides of copper, zinc and iron originating from the slider and the ring, and wear rates of both are comparable. In the regime of severe wear, the wear rate of the harder counterface is negligible, and the wear debris is metallic brass. There is extensive metallic contact, as shown by the low contact resistance.

Lubricants are employed to reduce both wear and the frictional force between surfaces: they act by introducing between the sliding surfaces a material with a lower shear strength than the surfaces themselves. In *hydrodynamic* lubrication the surfaces are separated by a fluid film, whereas in *boundary* lubrication the surfaces are separated by adsorbed molecular films.

The Reading List at the end of this Chapter will guide the reader in further reading on the subject of *tribology* – the study of friction, wear and lubrication – where the detailed mechanisms of these processes are fully discussed.

FURTHER READING

I.J. Polmear: *Light Alloys*, 3rd edition, Edward Arnold, London, 1995.

R.W.K. Honeycombe: Steels – Microstructure and Properties, 2nd edition, Edward Arnold, London, 1995.

D.T. Llewellyn: *Steels – Metallurgy & Applications*, Butterworth, London, 1992.

H.T. Angus: *Cast Iron: Physical and Engineering Properties*, Butterworth, London, 1976.

C.J. Thwaites: *Capillary Joining – Brazing and Soft-soldering*, Research Studies Press, Letchworth 1982.

Welding Handbook (Volume 2 – Welding Processes) Seventh Edition (ed. W.H. Kearns), American Welding Society, Miami, Florida (1978).

Kenneth Easterling: *Introduction to the Physical Metallurgy of Welding*, Butterworths, London, 1983.

Adhesives and Sealants, Engineered Materials Handbook Volume 3, ASM International, 1990.

I.M. Hutchings: *Tribology: Friction and Wear of Engineering Materials*, Edward Arnold, London 1992.

John M. West: *Basic Corrosion and Oxidation*, 2nd edition, Ellis Horwood, Chichester 1986.

4 Glasses and Ceramics

GLASSES

The random three-dimensional network of molecular chains present in a glass is illustrated diagrammatically in Fig. 1.22. The chains consist of complex units based on the SiO_4 tetrahedral unit. Pure silica can form a glass; it has a high softening temperature, so it is hard to work because its viscosity is high. Commercial glasses consist of silica mixed with metal oxides which greatly reduce the melting temperature, thus making the glass easier and cheaper to make. There are two important families of commercial glass:

Soda Lime Glasses

These have a typical composition (wt%)

$$70 \ SiO_2, \ 10 \ CaO, \ 15 \ Na_2O$$

and small amounts of other oxides. These glasses are employed for high volume products such as windows, bottles and jars. The added metal oxides act as *network modifiers* in the structure of Fig. 1.22. Thus when soda (NaO_2) is added to silica glass, each Na^+ ion becomes attached to an oxygen ion of a tetrahedron thereby reducing the cross-linking as indicated in Fig. 4.1. The effect of soda addition is thus to replace some of the covalent bonds between the tetrahedra with (non-directional) ionic bonds of lower energy. This reduces the viscosity of the melt, so that soda glass is easily worked at 700°C, whereas pure silica softens at about 1200°C. By the same token, this alloying of the glass to make it more workable reduces its strength at high temperature, so that silica glass must be used in applications requiring high temperature strength – such as the envelopes of quartz halogen lamps.

Soda lime glasses are also sensitive to *changes* in temperature, which, because of their large coefficient of thermal expansion ($\sim 8 \times 10^{-6} \ K^{-1}$) can develop high thermal stresses which can induce cracking. The second important family of glasses were developed to overcome this problem.

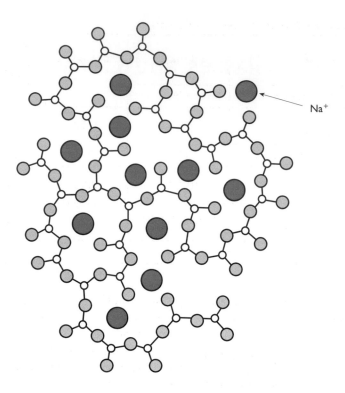

Na$^+$

Fig. 4.1 Sketch of the structure of soda lime glass.

Borosilicate Glasses

These have a typical composition (wt %)

$$80 \text{ SiO}_2, \; 15 \text{ B}_2\text{O}_3, \; 5 \text{ Na}_2\text{O}$$

and a small amount of Al_2O_3.

The coefficient of thermal expansion of these glasses is only one third of that of soda lime glass, and their resulting good thermal shock resistance makes them suitable for cooking and chemical glassware (Pyrex glass).

The properties of these glasses are compared in Table 4.1.

Thermal Shock Resistance

Fracture caused by sudden changes in temperature is a problem with glasses and ceramics. If a piece of the heated material is plunged into cold water, the surface layer immediately attains the water temperature before there is time for heat to flow from the interior of the specimen. The contraction of the surface layer is thus constrained, creating a biaxial surface stress given by:

$$\sigma = \frac{\alpha E \, \Delta T}{1 - \nu} \qquad (4.1)$$

Table 4.1 Some properties of glasses

Property	Units	Silica glass	Soda-lime glass	Boro-silicate glass	Glass ceramic
Modulus of rupture	MPa	70	50	55	>110
Compressive strength	MPa	1700	1000	1200	~1200
Young's modulus	GPa	70	74	65	92
Thermal expansion coefficient	10^{-6} K^{-1}	0.62	7.8	3.2	1.0
Thermal conductivity	W m^{-1} K^{-1}	1.8	1.8	1.5	2.4
Toughness (G_c)	J m^{-2}	~1	~1	~1	~10

where α is the thermal expansion coefficient, E is Young's modulus, ν is Poisson's ratio and ΔT the temperature difference between surface and interior of the specimen.

The value of ΔT_f required to generate the fracture stress of the material (σ_f) can be regarded as a merit index (R) for the thermal shock resistance of ceramics and glasses:

$$\Delta T_f = \frac{\sigma_f(1-\nu)}{\alpha E} = R$$

A specimen shape factor, S, may be included, giving:

$$\Delta T_f = RS$$

The value of the thermal expansion coefficient is thus of critical importance in this context, and it is obvious from the data in Table 4.1 why borosilicate glasses are superior in this respect to soda-lime glass.

This approach has made the implicit assumption that the surface temperature reaches its final value before there is any change in temperature in the bulk of the material – i.e. that an infinitely rapid quench has been applied. In practice, finite rates of heat transfer should be taken into account by considering the Biot modulus (β), defined by the equation:

$$\beta = \frac{r_m h}{K} \qquad (4.2)$$

where r_m is the radius of the sample, h the heat transfer coefficient and K the thermal conductivity.

If we define a non-dimensional stress σ^* as the fraction of stress that would result from infinitely rapid surface quenching, then

$$\sigma^* = \frac{\sigma}{\alpha\,E\,\Delta\,T/(1-\nu)} \qquad (4.3)$$

where σ is the actual stress observed.
But σ^* varies with time which has been found in terms of the Biot modulus to be:

$$\sigma^* = 0.31\beta \qquad (4.4)$$

A modified thermal shock resistance factor, R', can now be written

$$\Delta T_f = R'\,S\frac{1}{0.31\,r_m\,h}$$

where

$$R' = \frac{K\,\sigma_f\,(1-\nu)}{\alpha E} \qquad (4.5)$$

Thus, in addition to the factors mentioned earlier, good resistance to thermal stress failure also require a high thermal conductivity in the material.

Toughening of Glass

Glass shows very high strength in compression: its compressive breakage strength is over 1 GPa. This high value has been used in the design of submarine ocean research devices. These are glass spheres (filled with electronic instruments) which successfully withstand the high hydrostatic pressures near the ocean bed. The modulus of rupture of window glass is 50 MPa, and this weakness of glass in tension is due to the presence of small surface cracks ('Griffith cracks') arising from normal handling of the material. The toughness of glass (G_c) is only about $1\,\mathrm{Jm}^{-2}$, and its Young's modulus is 74 GPa, so that the Griffith equation (2.11) suggests that microcracks of the order $10\ \mu\mathrm{m}$ in depth are present.

In order to increase the strength of glass, these surface cracks must be prevented from spreading by ensuring that they do not experience a tensile stress. This is achieved by inducing surface compressive stresses into the glass, which can be done in two ways:

Thermal Toughening
The piece to be toughened is heated above its glass transition temperature, and the surface is rapidly chilled, for example

by a series of air-jets played upon it. The surface cools and contracts while the inside is still soft and at high temperature. Later, the inside cools, hardens and contracts, but the outer layers cannot 'give' and are thus compressed. In other words, the original temperature gradient, induced by the air-jets, is replaced by a stress gradient in the glass – the outer layers being in a state of compression, with a stress of about -100 MPa, the interior being in tension. Any internal strains influence the passage of polarized light through the glass, so that when one looks through a car windscreen while wearing polarizing spectacles it is possible to see the pattern of the air jets used in the toughening treatment.

Chemical Toughening

If the hot glass article is immersed in a molten salt such as potassium nitrate, some of the Na^+ ions in the surface of the glass will be exchanged with K^+ ions from the salt. K^+ ions are about 35% larger than Na^+ ions, and the time of treatment is chosen so that the ions diffuse to a depth of about 0.1 mm into the surface of the glass, which therefore attempts to occupy a greater volume. This is resisted by the material beneath the K^+-enriched surface which therefore exerts a compressive stress on the surface layers. Maximum stresses of -400 MPa can be achieved by this method, although the depth of the compressed layer is considerably less than with thermal toughening. Chemical toughening tends to be more costly than thermal toughening, but it can be used on thinner sections.

After either treatment, before a surface crack can be propagated the tensile stresses applied have to overcome these stresses of opposite sign, which result in a four- to ten-fold increase in strength. In other words, it is found that a sheet of toughened glass may be bent further before it will break, and furthermore the glass breaks into very small fragments which are much less dangerous that the sharp shards produced when annealed glass is broken. The toughened glass produces more cracks and therefore smaller fragments because it contains more stored elastic energy to propagate the cracks than in the case of annealed glass.

Environment-Assisted Cracking

Glasses, and many other oxide-based ceramics are susceptible to slow crack growth at room temperature in the presence of water or water vapour, if there is a surface tensile stress acting. This may lead to *time-dependent failure* of the specimen, and the effect can be expressed in terms of fracture mechanics, as discussed in Chapter 2.

Essentially, water vapour at the crack tip can react with the molecules and break the Si–O bonds by forming an hydroxide. When the crack has grown sufficiently for the critical stress intensity to be achieved, failure takes place – the phenomenon sometimes being referred to as 'static fatigue'.

GLASS CERAMICS

Glass is a Newtonian viscous solid, so it is easy to mould without introducing voids, but its high temperature strength is essentially low. A number of glass compositions have been identified which can be *crystallised* after the shaping process is complete. A controlled heat-treatment is required, first to nucleate and then to grow the crystals throughout the glass. The extent of crystallisation may exceed 90% by volume, and small crystal sizes of < 0.5 μm are produced in a glass matrix. The manufacture of glass ceramic products is characterized by the high-speed, high-volume economics of glass melting and forming, and provides material with new and often unique properties.

The vast majority of commercial glass ceramics are based on two aluminosilicate systems:

(a) β-Spodumene ($Li_2O.Al_2O_3.nSiO_2$), which is particularly useful for cookware and counter-top cooking surfaces because of its low thermal expansion ($\alpha = 1.10^{-6}$ K^{-1}). Commercial compositions utilize TiO_2 as a crystal nucleating agent, and inclusion of up to 0.5% fluorine in the composition leads to volatilization at the surface during the crystallisation process. This composition gradient in fluorine promotes the development of surface compressive stresses, with resulting strengths greater than 140 MPa.

(b) Cordierite ($2MgO.2Al_2O_3.5SiO_2$), which found applications as a radome or missile nose cone. The material combines high electrical resistivity with high mechanical strength and moderately low thermal expansion. The presence of 11 wt% cristobalite (SiO_2) in radome glass ceramic allows the material to be further strengthened by exposing it to an alkali solution which leaches the cristobalite phase from the surface, leaving a porous layer. This removes the mechanical surface damage due to grinding, and increases the strength from 120 MPa to 240 MPa.

CERAMIC MATERIALS

It is possible to classify ceramic materials into three groups:

1. Natural ceramics. Stone is the oldest of all constructional materials, and one may distinguish *Limestone* (largely $CaCO_3$) and *Granite* (aluminium silicates). Their mechanical properties are summarised in Table 4.2.

Table 4.2 Mechanical properties of some ceramics

Ceramic	Young's Modulus (GPa)	Compressive strength (MPa)	Modulus of rupture (MPa)
Limestone	63	30–80	20
Granite	60–80	65–150	23
Porcelain	70	350	45

2. Domestic ceramics. Pottery, porcelain, tiles and structural and refractory bricks are all in this category. These so-called vitreous ceramics are made from clays which are formed in the plastic state, they are then dried so they lose their plasticity and acquire enough strength to be handled for firing. Firing at temperatures between 800°C and 1200°C leads to a microstructure which on cooling consists of crystalline phases (mostly silicates) in a glassy matrix based on silica (Fig. 4.2). The structure contains up to 20% porosity, and

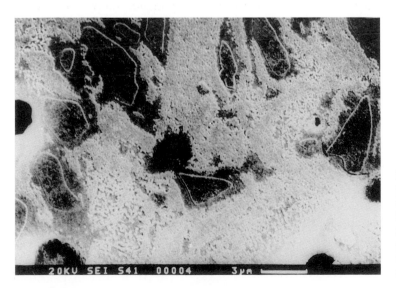

Fig. 4.2 SEM micrograph of porcelain specimen. Dark quartz crystals are outlined in glassy regions which surround them. The fine light grey crystals are of mullite (Courtesy Dr P. F. Messer and H. E. Okojie).

many microcracks, but a glaze may be applied to seal the surface pores.

The mechanical properties of porcelain are shown in Table 4.2.

3. *Engineering Ceramics*. A number of high performance engineering ceramics are now produced. These are essentially simple compounds such as oxides, carbides and nitrides in the pure crystalline state with very low, sometimes negligible porosity. In comparison with the traditional ceramics described above, they contain smaller microcracks, so their strength and toughness is improved, giving properties competitive with metals for applications such as cutting tools, dies and engine parts.

Processing of Modern Ceramics

Most ceramic fabrication processes begin with finely ground powder. Oxides such as alumina (Al_2O_3), magnesia (MgO) and zirconia (ZrO_2) occur naturally, but have to be purified by chemical processing before use as engineering ceramics. Silicon carbide (SiC) is manufactured by reacting SiO_2 sand with coke (C) at high temperature, and silicon nitride is also synthesised industrially – usually by reacting silicon powder with nitrogen at temperatures in the range $1250°C$ to $1400°C$. Prior to consolidation the powders are milled and graded into size, of diameter of the order 1 μm. They are then blended so that the subsequent shaping operation leads to material of optimum properties. The next stage is one of shape-forming, for which there are a number of possible processes.

Pressing requires the powder to be premixed with suitable organic binders and lubricants and preconsolidated so that it is free flowing. It is then compacted in a die to form small shapes such as crucibles, and insulating ceramics for electrical devices.

Slip casting is effected by suspending the ceramic particles in a liquid (usually water) and pouring the mixture into a porous mould (usually plaster) which removes the liquid and leaves a particulate compact in the mould. An organic binder is usually present in order that the casting has sufficient strength to permit its removal from the mould prior to the firing operation.

Plastic forming is possible if sufficient (25 to 50 vol%) of organic additive is present to achieve adequate plasticity. Injection moulding and extrusion may then be employed.

Strong, useful ceramic products are produced after the final densification by *sintering* at high temperature is carried out. Sintering brings about the removal of pores between the starting particles (accompanied by shrinkage of the component), combined with strong bonding between the adjacent particles. The primary mechanisms for transport are atomic diffusion and viscous flow. In some cases *hot die pressing* may be applied, whereby pressure and temperature are applied simultaneously, which may accelerate the kinetics of densification. Only a limited number of shapes can be produced by this technique, however.

The thermodynamic driving force for sintering is the reduction in surface energy (γ) by the elimination of voids. A spherical void of radius r will experience a closure pressure, P, given by:

$$P = -2\gamma/r$$

Thus the smaller the void, the greater the closure pressure. Efficient sintering is thus promoted by the use of precursor powders of fine particle size. The diffusional process requires the presence of lattice vacancies, and ceramics of covalent bonding have a very high formation energy for vacancies, and so exhibit low solid state diffusion rates, giving poor densification properties. Atom transport is predominantly by grain boundary diffusion, so again a fine grain-size is essential for efficient densification. In most cases a 'densification aid' or grain growth inhibitor is added to the ceramic to achieve maximum density and minimum grain size. In the case of carbide and nitride ceramics these are metal oxides; LiF is added to alumina and magnesia for this purpose. The additives segregate to the newly-formed grain boundaries during sintering and increase the diffusion coefficient by forming low melting-point or low viscosity glass phase. The best sintering aids also suppress *grain growth* in the component, which would otherwise lead to a reduction in the number of diffusion paths, thus slowing the densification rate. Voids would become 'stranded' in large grains, with no fast pathway for mass transfer, and remain as a likely source of cracking when the component is under stress in service.

The microstructure of a pure, polycrystalline engineering ceramic can be seen by polishing, etching and magnifying. The important features are the grain size and the degree of porosity, and a dense ceramic is thus similar in microstructure to a polycrystalline metal (Fig. 4.3). Table 4.3 summarises the properties of some of these materials. The data give a general indication of typical strengths, but test data from current suppliers should be used for analytical design or life prediction calculations.

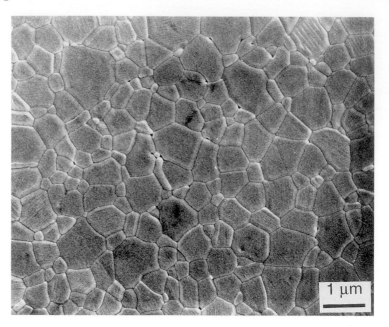

1 µm

Fig. 4.3 Pure polycrystalline alumina (courtesy of Dr F. L. Riley and Dr M. Miranda-Martinez)

Table 4.3 Properties of some engineering ceramics

Property	Units	Alumina	Zirconia	Silicon nitride (RBSN)	Silicon carbide (RSSC)
Modulus of rupture	MPa	300–400	200–500	200–350	450
Compressive strength	MPa	3000	2000	2000	2000
Young's modulus	GPa	380	138	150–180	400
Thermal expansion coefficient	10^{-6} K^{-1}	8.5	8	2.6	4.5
Thermal conductivity	W m^{-1} K^{-1}	25.6	1.5	12.5	100
Toughness (G_c)	J m^{-2}	25	80	10	25

Alumina (Al_2O_3)

Many hundreds of tons of alumina powder is produced annually from the mineral bauxite, and it is used in the manufacture of porcelain, crucibles, wear-resistant parts such as cutting tools and grinding wheels, medical components and a variety of other components. It forms ionic crystals of hexagonal structure, Fig. 4.4, with close-packed layers of oxygen ions, with the Al^{3+} ions occupying interstices such that each is surrounded by six O^{2-} ions. One third of the Al^{3+}

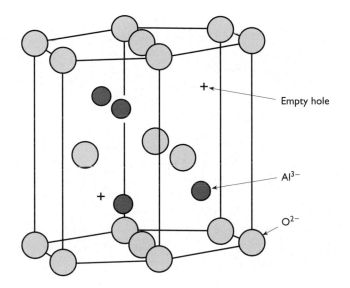

Fig. 4.4 The crystal structure of alumina.

sites remain empty, so that the overall ionic charges are in balance.

Reducing the grain size increases both the fracture strength and the toughness of alumina (Fig. 4.5). The manufacturing conditions for fine grain size (low temperatures, short time) are in conflict for those to minimise porosity (high temperatures, long times), and a compromise has to be made by using sub-micron powder particles as a starting material, and by the addition of 0.05–0.2% magnesia (MgO) which prevents

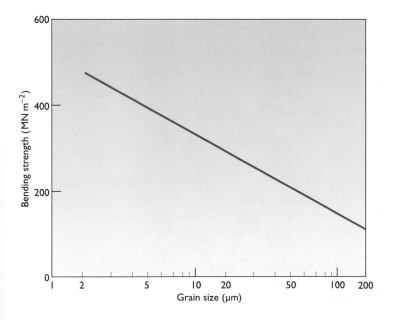

Fig. 4.5 Showing increase in strength with decreasing grain size in alumina.

grain growth by 'pinning' the alumina grain boundaries. Hot pressing at 1350–1800K may enhance sintering, so that a shorter time is required at the sintering temperature of 1850–2000K.

Zirconia (ZrO₂)

Zirconia (ZrO_2) is an engineering ceramic of growing importance: it is again an ionically bonded material, and it exhibits three distinct crystal phases. Above 2300°C it is cubic, between 1150° and 2300°C it is tetragonal and below 1150°C it has a monoclinic structure. The cubic form consists of zirconium ions on a face-centred cubic lattice (Fig. 4.6) with oxygen ions occupying certain 'holes' in the structure. ZrO_2 undergoes a 3.5% volume expansion during cooling below 1000°C due to its change in crystal structure to monoclinic, and this causes catastrophic failure of any part made of pure polycrystalline zirconia. Addition of CaO, MgO or Y_2O_3 to the zirconia results in a cubic crystal structure that is stable over the complete temperature range and does not undergo a phase transformation. This is referred to a *stabilised zirconia*.

Stabilised zirconia has a low fracture toughness and a poor resistance to impact. By not adding enough CaO, MgO or Y_2O_3 to stabilize the ZrO_2 completely and by careful control of processing, mixtures of the stabilized cubic phase and the metastable tetragonal phase that have very high fracture toughness are achieved. This type of material is referred to a *partially stabilised zirconia (PSZ)*.

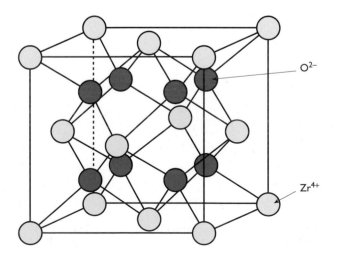

Fig. 4.6 The cubic crystal structure of ziroconia.

Transformation Toughening

A suitable microstructure consists of a matrix of cubic zirconia containing a dispersion of particles of metastable tetragonal zirconia. The transformation of the small zirconia particles from tetragonal to monoclinic zirconia is inhibited by the elastic constraint of the surrounding matrix. Ahead of a propagating crack in such a material there is a dilatational stress field; this interacts with the constraining stress field around a metastable particle and initiates transformation. Transformation will occur to some distance within the stress field, and thus behind the crack tip there will be a wake or process zone of transformed particles (Fig. 4.7). The volume expansion of these particles acts as a crack closure strain and thus reduces the stress intensity at the crack tip. This means that a further stress has to be imposed to continue crack propagation, so the failure stress (and hence the toughness) increases. PSZ zirconia can have fracture strengths of about 600 MPa with fracture toughnesses of around 8–9 MPa m$^{\frac{1}{2}}$.

Zirconia has other properties which make it a very interesting engineering ceramic. As seen in Table 4.3, it has a very low coefficient of thermal conductivity (1.5 W m^{-1} K^{-1}, compared with 25.6 W m^{-1} K^{-1} in the case of alumina) together with a very high thermal expansion coefficient (8×10^{-6} K^{-1}) – two or three times that of most ceramics and almost the same as cast iron or steel. This makes zirconia

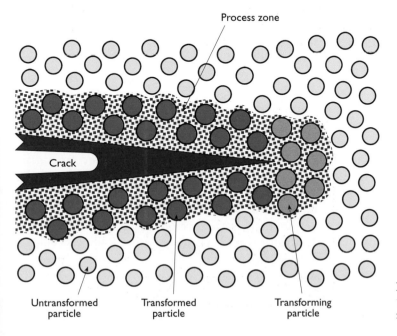

Process zone

Crack

| Untransformed particle | Transformed particle | Transforming particle |

Fig. 4.7 Mechanism of transformation toughening in zirconia ceramic.

a candidate for insulating engine components, since any coatings will not have the severe problems of thermal expansion mismatch found with other non-metallic surface layers.

Nitride Ceramics

Silicon Nitride (Si_3N_4)

Silicon nitride (Si_3N_4) is a covalently bonded crystalline ceramic. Its crystal structure is based on the packing of tetrahedra: each tetrahedron has a silicon atom at the centre and a nitrogen atom at each corner, just like the SiO_4 tetrahedra in silica. In silicon nitride, a nitrogen atom is shared by *three* tetrahedra (in contrast to the oxygen sharing *two* tetrahedra in silica) forming crystals of hexagonal symmetry. The tetrahedra are thus held more rigidly in silicon nitride than in silica, and less variation in bond length and angle is allowed, leading to its stronger and stiffer mechanical properties.

It is strong, hard, wear-resistant, stable up to 1800°C and oxidation-resistant. With its low thermal expansion coefficient it has excellent resistance to thermal shock, and has application in load-bearing components such as high-temperature bearings and engine parts. The diffusivity of silicon nitride is very low, and the chief difficulty is in fabricating fully dense material.

Reaction Bonded Silicon Nitride

Reaction Bonded Silicon Nitride (RBSN) is fabricated from silicon powder. The silicon powder is processed to the desired shape by pressing, slip casting, or other suitable process and then placed in a furnace under a nitrogen atmosphere and heated to approximately 1400°C. The reaction:

$$3\ Si + 2\ N_2 \rightarrow Si_3N_4$$

takes place. Approximately 60% weight gain occurs during nitriding, but less than 0.1% dimensional change, so that excellent dimensional control is possible upon the finished product. Because of the need to allow access of nitrogen during processing, the structure of the product will be that of an interconnecting network of voids, and RBSN is typically only 70–80% of full density. Because no grain boundary glassy phases are present, RBSN does not suffer rapid degradation of strength at elevated temperature, and it is useful as a static component in high temperature applications.

Sialons

Silicon nitride can be 'alloyed' with aluminium oxide because the AlO_4 tetrahedron is similar in size to the SiN_4 tetrahedra in silicon nitride. It is not surprising, therefore, that up to two-thirds of the silicon can be replaced by aluminium as long as sufficient nitrogen is replaced by oxygen to preserve charge neutrality.

Sialon is much more resistant to oxidation than silicon nitride, because a surface protective film of 'mullite' $(3Al_2O_3.2SiO_2)$ is developed rather than the weaker coat of silica formed upon RBSN.

A high density is achieved by the use of yttria or yttria plus alumina as densifying agents in sintering, by the formation of a yttria-sialon glass in the grain boundaries. This glass may subsequently be crystallised in the grain boundaries of the sialon, so that these materials then have very good creep properties.

Carbide Ceramics

Silicon Carbide

Reaction-sintered silicon carbide (RSSC) is prepared from a mixture of graphite and silicon carbide powders, which is infiltrated with molten silicon at temperatures $> 1720K$. The process is conducted in a nitrogen atmosphere, and the product contains little porosity, although there is a proportion of residual silicon in the microstructure, which limits the high temperature strength of the material. Strengths superior to RBSN may be achieved, although considerable scatter is strength is found.

Sintered silicon carbide (SSC) is prepared from fine powders, using boron and carbon as sintering aids. Properties are marginally better than for RSSC, although the growth of coarse (> 50 μm) grains during sintering are a strength-limiting factor.

Cement and Concrete

Concrete is a composite material made up of a matrix of cement into which is embedded particles of sand and aggregate: the cement provides a means of binding together low cast sand and gravel into a product of engineering value. The mix must contain enough cement to coat all the aggregate particles and to fill the spaces between them. The aggregate should consist of a range of gravel particle sizes, so that the smaller particles help to fill the spaces between the larger

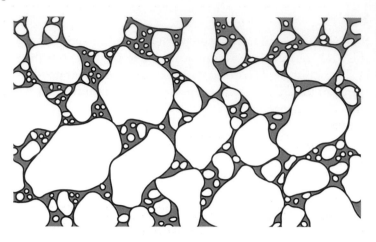

Fig. 4.8 Graded aggregate in a matrix of cement.

particles, thus reducing the amount of cement required, Fig. 4.8.

Portland Cement (OPC) is the most important of the binders used in concrete. It was patented in the 1820s, and so named because its colour resembles that of the natural limestone found at Portland in Dorset. It is made by firing a suitable mixture of clay and chalk, and then grinding the product to a fine powder. It consists of a mixture of particles of ionically bonded solids; its constitution (in weight %) is as follows, with a contracted symbol shown in brackets to represent some of the compounds:

55% tricalcium silicate (**C₃S**)
20% dicalcium silicate (**C₂S**)
12% tricalcium aluminate (**C₃A**)
8% tetracalcium aluminoferrite (**C₄AF**)
3.5% hydrated calcium sulphate (gypsum) (**C₄S̄H₂**)
1.5% minor constituents, CaO (**C**), K₂O, Na₂O.

The shorthand symbols have the following meaning:

Oxide	CaO	SiO₂	Al₂O₃	Fe₂O₃	H₂O	SO₃
Symbol	**C**	**S**	**A**	**F**	**H**	**S̄**

Setting and Hardening

After water is added to the cement powder, the resulting *cement paste* can be shaped and used for a period approaching two hours. Figure 4.9 is a calorimetric curve showing the heat evolution with time during the hydration of Portland cement. During the period marked Stage I, a considerable amount of heat is evolved (a peak of 200 W kg^{-1} after 30 s), which constitutes the stage of *setting*. Setting ends when *hardening* starts and the cement paste begins to solidify (Stage II).

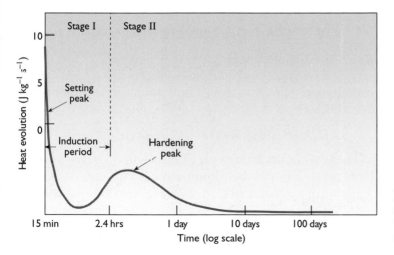
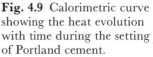

Fig. 4.9 Calorimetric curve showing the heat evolution with time during the setting of Portland cement.

Figure 4.10 shows the change in compressive strength with time for Portland cement: hardening is seen to start after an hour or so, and continues at a decreasing rate for a number of years.

The Setting Reactions

The setting reactions can occur over a long period of time, as the different constituents of the cement react at different rates – the rate of hydration is in the approximate order:

$$C_3A > C_3S > C_4AF > C_2S$$

The rapid hydration of C_3A would lead to 'flash setting' of the cement, due to the formation of calcium aluminate

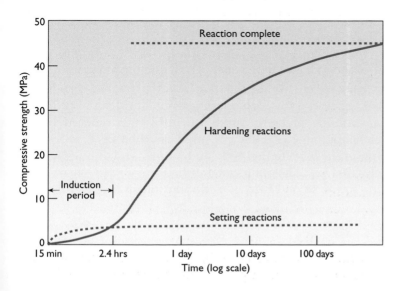
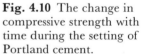

Fig. 4.10 The change in compressive strength with time during the setting of Portland cement.

hydrates:

$$C_3A + 21H \rightarrow C_4AH_{13} + C_2AH_8$$

which then convert readily to other hydrates.

This is avoided by the addition of gypsum $(C_4\bar{S}H_2)$ which combines with C_3A to form sulpho-aluminate hydrates such as ettringite:

$$C_3A + C_4\bar{S}H_2 + 26H \rightarrow C_6A\bar{S}H_{32} \text{ (ettringite)}$$

These are in the form of very fine needle-like crystals which form slowly around the aluminate, and flash set is avoided. The rate of hydration of C_3A then becomes comparable with that of C_3S, and it is these hydration reactions proceeding together which are mainly responsible for the initial high rate of heat evolution in the cement paste.

The hydration reactions of the two calcium silicates are very similar:

$$2C_3S + 7H \rightarrow C_3S_2H_4 + 3CH$$

$$2C_2S + 5H \rightarrow C_3S_2H_4 + 3CH$$

$C_3S_2H_4$ (calcium silicate hydrate) is referred to as C–S–H gel, and it forms as a poorly crystalline material in the size range of colloidal matter (less than 1 μm in any dimension). The surfaces of the cement grains become coated with silicate and aluminate gels, which causes the drastic deceleration of the reaction rate. The cohesion of the cement paste during the setting period is due to gel-to-gel contact on adjoining grains, aided to an increasing extent by the interlocking of the ettringite crystals, and indicated in the diagram of Fig. 4.11a.

C_4AF acts as a flux in the cement kiln, but its hydration products play little role in the setting and hardening processes. It does, however, contribute colour to the cement.

The hardening process starts after about 3 hours. The envelope of gel around the cement grains also acts as a semipermeable membrane for water. Water is drawn through the gel because of the high concentration of calcium inside, and an osmotic pressure builds up until it is relieved by the envelope bursting. The ruptured gel then peels away from the grain forming gel foils and fibrils, and tubules of C_3A (Fig. 4.11b). The grain can now hydrate further, corresponding to the second heat of evolution shown in Fig. 4.9, and a

Fig. 4.11 Illustrating the setting and hardening of Portland cement, showing (a) gel-to-gel contact, (b) growth and interlocking of fibres, and (c) growth and interlocking of fibres and crystals.

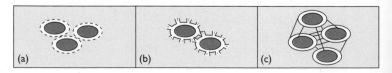

multitude of fibres grow and multiply. Hardening corresponds to the interlocking of these fibres, causing the paste to become rigid. Crystalline plates of $Ca(OH)_2$ (**CH**), known as *portlandite* also form, and hardening continues as further fibres and crystals form (Fig. 4.11c).

Control of the Rate of Hydration of Cement
The rate of cement hydration is affected by the ambient temperature. In cold conditions, *accelerators* (e.g. calcium chloride) are added to concrete to increase the kinetics of hydration, and may thus, for example, allow the cement to set before frost can damage it. In hot environments a *retarder* is used (e.g. sugar) to reduce the setting speed by inhibiting the growth of **C-S-H** gel. The use of such additives enables the setting and hardening of the cement to be controlled so that the use of the cement can be extended to a wider range of applications. If it is essential for the cement to set quickly and to exhibit a rapid initial gain in strength, the kinetics can be increased as above, whereas if a cement is required to set slowly, it can be slowed down by the appropriate additive.

Structure/Property Relationships in Cement and Concrete

The Elastic Modulus
Concrete is a composite material (Fig. 4.8). The Young's modulus of gravel is three or four times that of cement, and one may calculate the value of the modulus of a concrete mixture from the relation:

$$E_{concrete} = \left[\frac{V_a}{E_a} + \frac{V_c}{E_c} \right]^{-1}$$

where V_a and V_c are the volume fractions of aggregate and cement, and E_a and E_c are their moduli, with typical values 130 GPa and 32 GPa respectively. V_a can range from 0.45 to 0.75, giving moduli for such concretes of the order 50 GPa.

The Effect of Porosity
We have seen that the strength of concrete develops over a period of time, and data are only significant when related to the time after casting. The 28-day strength is often used as a standard parameter. Strength is commonly described by subjecting a specimen to a compressive axial load and recording the crushing strength, although for other applications it may be tested in flexure.

The modulus of rupture so obtained is only about 10% of its compressive strength, the reason for this being the presence

Fig. 4.12a Optical micrograph of hydrated Portland cement showing porosity, ×25, (Courtesy Dr G.W. Groves).

of *porosity* , Fig. 4.12a. The pore size distribution is very wide, with different sized pores having a different effect on the strength of the cement. The smallest pores, gel pores, are < 10 nm in size, and form within the **C–S–H** gel. Since they are so small, these pores do not have a large effect on the strength of the cement. Capillary pores are > 10 nm, and can be found between the fibres of **C–S–H** outer product to form a three-dimensional pore network. Since these are larger, they will cause some reduction in the strength of the cement. The major problem with porosity, however, arises from air bubbles resulting from poor processing, or pockets of water. These can be up to 1 mm in size and can propagate under stress leading to premature failure of the component, in accordance with the Griffith equation:

$$\sigma = (2E\gamma/\pi c)^{\frac{1}{2}} \qquad (2.10)$$

The porosity is essentially dependent upon the water/cement ratio, an increase in which causes the strength of the cement to decrease. As the amount of water increases above that necessary for the complete hydration of the cement mixture, it produces a more porous structure, resulting in a decrease in strength (Fig. 4.12b). The amount of air in the concrete also affects strength, but to a lesser extent.

In compression a single large flaw is not fatal (as it is in tension). In a compression test, cracks inclined to the compression axis experience a shear stress which causes them to propagate stably. They twist out of their original orientation to propagate *parallel to the compression axis*. They eventually

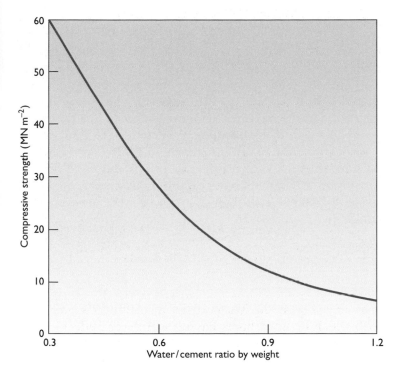

Fig. 4.12b Compressive strength vs water/cement ratio.

interlink to form a *crush zone* which develops at an angle of 30° to the compression axis. The form of the resultant stress–strain curve is as shown in Fig. 4.13.

The Effect of the Aggregate

The size and shape of the aggregate and the aggregate/cement ratio all affect the strength of concrete. A mix

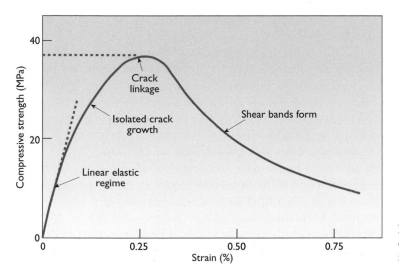

Fig. 4.13 The stress–strain curve for cement or concrete in compression.

containing large particles of aggregate will have a higher strength than one containing small aggregate. Furthermore, a coarse crushed rock aggregate will have greater strength than a smooth, rounded aggregate, since in the latter situation shearing tends to take place round the aggregate rather than through it. Similarly, the greater the proportion of aggregate in the mixture, the more probable it is that shear occurs through it. Unfortunately, cement with a low water/cement ratio and a high proportion of coarse aggregate will be almost unworkable, although this is the condition for maximum strength. The composition of a practical high strength concrete will thus be a compromise between these conflicting requirements.

High Strength Cements

The Pozzolanic Reaction

Pozzolanas are materials which contain constituents that will combine with lime in the presence of water to form stable insoluble compounds possessing cementing properties. This is, in effect, the pozzolanic reaction:

$$\text{lime} + \text{pozzolana} + \text{water} = \textbf{C–S–H}$$
$$\textbf{C} + \textbf{S} \qquad\qquad + \textbf{H} = \textbf{C–S–H}$$

Pozzolanas may be natural, generally volcanic ash containing silica-rich glass, or artificial, such as pulverized fuel ash and burnt clays. Another source of relatively pure silica is 'condensed silica fume', which is a by-product of the production of silicon by the reduction of quartz in an electric furnace.

Up to 25% of microsilica may be added to cement. The ultrafine (~ 0.2 μm) particles act as a 'space filler' between particles, reducing porosity and consequently yielding a marked increase in strength. Sufficient time has to be allowed for the increased strength to be apparent, however, since the reactions involved are quite slow.

The pozzolanic reaction has reduced heat evolution, thus making it suitable for use in concrete structure of large dimensions. Pozzolanic additions to OPC will thus reduce its tendency to cracking, due to a reduced thermal contraction in the mass and an increased ability to creep under load. Pozzolanas are added to mortars, cements and concretes to increase their durability and economy, and they also give a cement that can set under water. They are used for concrete in marine, hydraulic and underground structures.

Superplasticisers

These additions allow a given degree of workability to be obtained at a reduced water/cement ratio. The ratio may be reduced by up to 20% in the presence of, for example, sulphonated naphthalene formaldehyde condensate, which is absorbed at the solid/liquid interface. They have the effect of neutralising the surface attractions between individual particles of the cement, causing them to form a less open structure, which therefore requires less water to fill spaces and provide fluidity to the paste. By decreasing the water/cement ratio in this way, superplasticisers act to increase the strength of the cement product.

By the addition of both superplasticisers and microsilica, cements with a six-fold increase in tensile strength have been achieved.

Macrodefect-Free (MDF) Cement

Recently a method for eliminating larger pores in cement has been used to produce MDF cement with a maximum pore size of ~ 15 μm. About 5% of a suitable polymer (e.g. polyacrylamide gel) is added to the cement paste, prepared from two size fractions of cement powder (~ 5 μm and 75–125 μm) to give denser packing. This is subjected to a dough-type mixing, followed by a compression at 5 MPa. This has the effect of eliminating the air, and the resulting cement, with < 1% porosity, exhibits a flexural strength of over 100 MPa, compared with 7 MPa for normal cement.

Reinforced Concrete and Pre-Stressed Concrete

Concrete which is used for structural members which experience tensile stresses, such as beams, must be reinforced so that the whole of the tensile load is taken by the reinforcement. It is common practice to use bars of heavily cold-worked mild steel of diameter 16–40 mm, deformed during manufacture to form surface protrusions, which can aid the interfacial bonding between steel and concrete. As the concrete sets, it shrinks and grips the steel, so that the tensile strength of the composite can be put equal to that of the reinforcement in the direction of the load. Steel and concrete happen to possess very similar coefficients of thermal expansion, so that changes in temperature do not generate interfacial stresses. The alkaline environment provided by the concrete surrounding the steel acts as a corrosion inhibitor for the steel.

Under tensile loading, assuming 'no slip' at the interface, the concrete and the steel undergo the same strain. Since the fracture strain of concrete is less than that of steel, extensive cracking occurs in the vicinity of the concrete/steel interface on the side of the beam in tension. It is important to limit the stress in the steel so that these cracks do not extend to the outside surface of the beam.

The effectiveness of the reinforcement can be increased by elastically *prestressing* the high tensile steel bars in tension. When the concrete has hardened, becoming anchored to the steel, the load on the steel is released bringing the concrete into a state of compression. The magnitude of the precompression should be similar in magnitude to that of the anticipated tensile stress in the beam. When the reinforced beam is subjected to tension in service the effect is to unload the precompression, and the concrete should not experience a tensile stress.

In 'post-tensioning', ducts are left when the structure is cast, and wires are threaded loosely in the ducts. When the concrete sets the wires are tensioned and anchored to the ends of the ducts. Liquid grout is then injected to fill the ducts. This method of stressing may be used in the construction of bridges, since a long beam can be manufactured from short segments, and then tensioned in this manner after assembly.

Durability of Concrete

Lack of durability due to chemical causes (as opposed to mechanical causes such as abrasion) arises from attack by sulphates, acids, sea water and other chlorides. Since this attack takes places *within* the concrete, the degree of permeability of the concrete is critical. The permeability is normally governed by the porosity of the cement paste, which in turn is determined by the water/cement ratio employed, and by the degree of hydration.

If the concrete is sufficiently permeable that agressive salts can penetrate right up to the reinforcement, then corrosion of the reinforcement will take place, if water and oxygen are also available, leading to possible cracking of the structure.

The development of cracks in concrete may also arise from volume changes due to shrinkage and temperature variations. In practice these movements are restrained, and therefore they induce stress. If these stresses are tensile in character then cracks may arise, since concrete is so weak in tension.

In addition to shrinkage upon drying, concrete undergoes carbonation shrinkage. Carbon dioxide (CO_2) is of course present in the atmosphere, and in the presence of moisture it forms carbonic acid which reacts with **CH** to form $CaCO_3$; other cement compounds are also decomposed. Carbonisation proceeds extremely slowly from the surface of the concrete inwards, at a rate depending on the permeability of the concrete, and its moisture content. Carbonisation neutralises the alkaline nature of the hydrated cement paste and thus the protection of steel reinforcement from corrosion. Thus if the full depth of concrete covering the reinforcement is carbonated, and moisture and oxygen can penetrate, again corrosion of the steel and possibly cracking will result. However, by specifying a suitable concrete mix for a given environment, it is possible to ensure that the rate of advance of carbonisation declines within a short time to a value of less than 1 mm per year. Therefore, provided an adequate depth of cover is present (as specified in British Standard BS 8110:1985), the passivity of the steel reinforcement can be preserved for the design life of the structure.

READING LIST

H. Rawson: *Glasses and their Applications*, The Institute of Materials, London, 1991.

P.W. McMillan: *Glass-Ceramics*, 2nd Edition, Academic Press, London, 1979.

David W. Richerson: *Modern Ceramic Engineering*, 2nd edition, Dekker, New York, USA 1992.

I. Soroka: *Portland cement paste and concrete*, Chemical Publishing Co., New York, 1980.

A.M. Neville and J.J. Brooks: *Concrete Technology*, Longmans, Harlow, 1987.

General Reference

Concise Encyclopedia of Advanced Ceramic Materials, R.J. Brook ed. Oxford: Pergamon Press, 1991.

Engineered Materials Handbook Volume 4: Ceramics and Glasses, Ohio, ASM International 1991.

5 Organic Polymeric Materials

In Chapter 1 we classified organic polymers as **thermoplastics** and **thermosets**. The overriding consideration in the selection of a given polymer is whether or not the material can be processed into the required article easily and economically.

FORMING PROCESSES FOR POLYMERS

Thermosets are heated, formed and cured simultaneously, usually by compression moulding, Fig. 5.1. The product can be removed from the mould while still hot, so the cycle time can be relatively short.

Thermoplastics soften when heated, and may be pumped, mixed and then shaped by one of several possible methods.

Extruder-Based Processes (Fig. 5.2)

The simplest and most common way of achieving extrusion is to use a single Archimedean screw in a heated barrel. Granules of polymer are fed into the screw, which compacts and mixes it and transports it to the heated region where melting occurs.

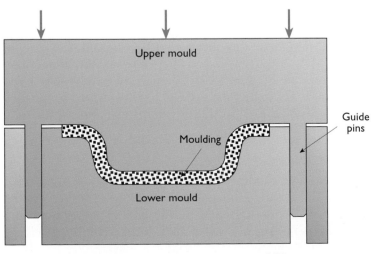

Fig. 5.1 Compression moulding.

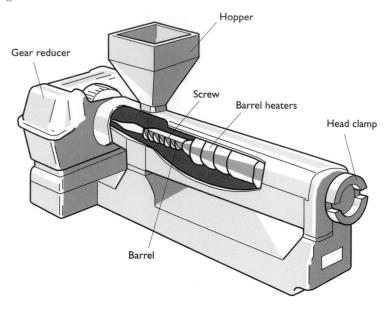

Fig. 5.2 Schematic view of a single screw extruder.

When the temperature is in the region of 1.3 to 1.6 T_g, the melt is finally forced through a die and then cooled to give tubes, sheet, ribbon or rod. Post-die devices which draw down or increase the dimensions of emerging simple shapes are widely used. For example, air can be blown into a formed tube to expand it to several times its diameter.

The deformation and flow behaviour of a given polymer (known as its *rheological properties*) are thus of paramount importance, and considerable emphasis is given by engineers to the measurement, presentation and use of rheological data for polymer melts.

Injection Moulding

Here the emerging melt from a screw extrusion device is injected under pressure into a cold split mould. Pressures up to 120 MPa are required, and mouldings of high dimensional precision may be produced, although the process has a slow cycle time since the product has to cool before its removal from the mould in order that its shape is maintained.

Blow Moulding

In the blow moulding of plastic bottles, melt is extruded out of an annular die to form a parison. Air is then injected into the centre of the annulus and a cold mould closes around it (Fig. 5.3). The top of the parison is snipped off by the closing

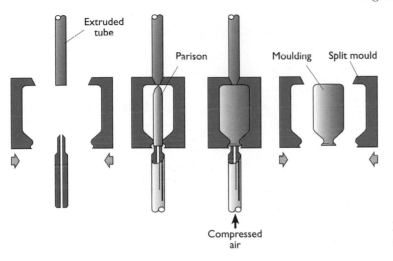

Fig. 5.3 The blow moulding of plastic bottles.

of the mould and the bottom becomes the mouth of the bottle. Air is injected and the inflated parison solidifies in the mould which then opens and ejects the bottle, and the whole cycle is repeated.

Directionality of Properties

The methods used to fabricate polymers have a strong influence on the properties and performance of the product. The variations in properties are largely associated with the differences in molecular orientation in the manufactured object. Molecules in the melted polymer flow in a non-Newtonian manner and tend to line up in the direction of flow. Molecules in thermoplastic polymers are much longer than those in the unreacted liquid resin precursors to thermosets.

The orientation effect in the injection moulding of thermoplastics is frozen as the melt solidifies upon cooling. The resulting tensile strength in the direction of molecular alignment is much greater than that in the cross flow direction.

MECHANICAL PROPERTIES

We will categorise the mechanical properties of polymers by *stiffness*, *strength*, and *toughness*. Figure 5.4 shows schematic stress–strain curves obtained at room temperature for the main classes of polymer. Curve 1 shows long range elasticity typical of the **elastomers**. These are wholly amorphous polymers, but the chains are lightly cross-linked so that the large extensions are wholly recoverable.

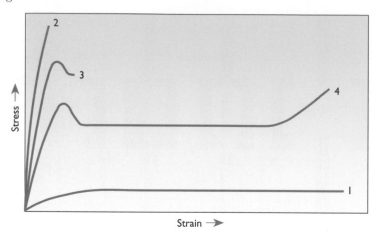

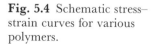

Fig. 5.4 Schematic stress–strain curves for various polymers.

Curve 2 is typical of polystyrene below its T_g, showing near linear elastic behaviour up to brittle fracture. Curve 3 is typified by (amorphous) polycarbonate, which yields in a ductile manner, exhibiting tough behaviour.

We will now consider the mechanical properties in more detail. We have so far assumed that the mechanical properties of polymers are independent of the *rate* of testing, which is not the case because the properties are governed by the mobility of the constituent polymer chains. This mobility is in turn dependent on the inherent stiffness of the chain, the degree of chain entanglement, the extent of crosslinking and the degree of crystallinity.

Stiffness

Consider an individual molecule in an amorphous polymer. When a stress is applied, deformation can take place by two processes, *bond stretching and bond angle opening,* and *rotation of segments of chain about the chain backbone.* Below the glass transition temperature the former are the main deformation mechanisms. As the temperature increases above T_g, however, individual backbone bonds are able to rotate (Fig. 5.5), and since the carbon-carbon bonds are at an angle of $109°\ 28'$ to one another, this ability to rotate can bring about enormous shape changes in the polymer chain, which becomes randomly kinked. Additionally, individual chains can slide locally relative to each other, with other regions remaining elastically deformed. On unloading, these elastic regions pull the polymer back to its original shape. This viscous process takes time, and the polymer will exhibit *leathery* properties whose response can be modelled empirically by 'spring-and-dashpot' combinations as described in Chapter 2.

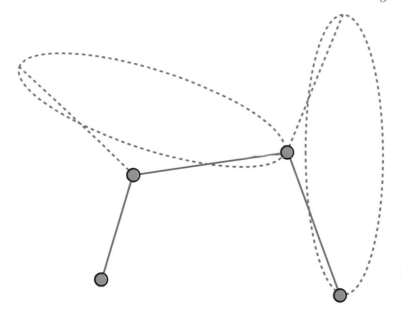

Fig. 5.5 Rotations about bonds in a polymer chain.

The mechanical properties of polymers are therefore **time and temperature dependent**, and Fig. 5.6 illustrates schematically how, *for a constant loading time*, the elastic modulus of a polymer will change with the temperature of deformation normalised with respect to T_g. At temperatures well below T_g it will be brittle-elastic, then passing through the T_g range it becomes visco-elastic, then rubbery and eventually viscous. As indicated in Fig. 5.6, the modulus can change by a factor of 10^3 or more over this range.

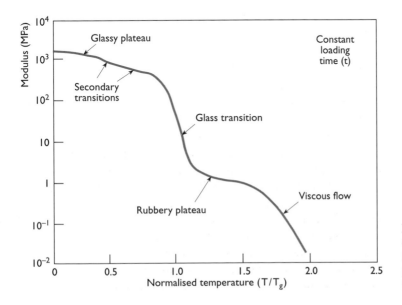

Fig. 5.6 Showing for a constant loading time how the elastic modulus of a polymer changes with temperature.

Rubber Elasticity

Curve 1 of Fig. 5.4 is typical of elastomers above their T_g, showing a low modulus but, because of cross-linking between the chains, exhibiting a reversible elasticity up to strains of several hundred per cent. In the unstressed state, the molecules are randomly kinked between the points of cross-linkage. On straining, the random kinking is eliminated because the polymer chains become aligned, and at high strains x-ray diffraction photographs of stretched rubber show that this alignment gives rise to images typical of *crystallinity*.

Thermodynamically, the deformation of an elastomer may be regarded as analogous to the compression of an ideal gas. Applying a combination the first and second laws of thermodynamics to tensile strains, we may write:

$$dU = TdS + FdL - PdV \qquad (5.1)$$

where F is the tensile force and L the length of the specimen. In elastomers Poisson's ratio is found to be approximately $\frac{1}{2}$, which implies that the tensile elongation causes no change in volume. So if an ideal elastomer is extended *isothermally*, then $dU = 0$ and equation 5.1 gives

$$F = -T\left(\frac{\partial S}{\partial L}\right)_{T,V} \qquad (5.2)$$

The ordering of the molecules by the applied strain *decreases* the entropy of the specimen. The strained state is thus energetically unfavourable and when the specimen is released it will return to a higher-entropy state where the chains have random conformations.

Treloar has calculated $(\partial S/\partial L)_T$ for an isolated long-chain molecule and for a model network of randomly kinked chains cross-linked at various points. If p is the probability of a particular configuration in a irregularly-kinked chain, then the entropy of the chain can be written:

$$S = k \log p$$

where k is Boltzmann's constant.

Treloar shows that $S = \text{constant} - kb^2L^2$, where b is a constant related to the most probable linear distance between the ends of the chain. So from equation 5.2, the tensile force required to extend by dL a chain whose ends are separated by a distance L is:

$$F = 2kTb^2L$$

The force is thus proportional to the temperature, which is in good agreement with experiment in that elastomers differ from crystalline solids, where the elastic moduli decrease as the temperature rises.

Elastomers do not show the Hookean behaviour of $F \propto L$, however, as seen in Fig. 5.4, curve 1, and Treloar extended his single-chain model to the case of a cross-linked network to give an expression of the form:

$$F = NkT(\lambda^2 - \lambda^{-1})$$

where N is the number of separate chains per unit volume, and $\lambda = L/L_o$.

The measured stress–strain curves of elastomers are in good agreement with this relationship for strains of up to 400%.

In simple shear, the shear strain γ is equal to $(\lambda - \lambda^{-1})$, and the force–deformation relationship is

$$F = NkT\gamma$$

so that Hooke's Law is obeyed in simple shear, and the shear modulus in the unstrained state is:

$$G = NkT$$

showing that the shear modulus of elastomers will increase as the number of cross-links between the polymer chains is increased.

Polymers With Higher Tensile Modulus

The tensile modulus of polymers is increased if the molecular chains are cross-linked together. For example, cross-linked rigid thermosets were among the first synthetic polymers to be manufactured. Such materials exhibit the highest tensile moduli (and lowest breaking strains) of common polymers, typified by curve 2 in Fig. 5.4.

Yielding

Drawing

Many partially crystalline polymers, such as polyethylene, polypropylene and polyamide show a tensile response illustrated in curve 4, Fig. 5.4. A non-linear pre-yield behaviour is followed by a yield point. The stress then drops and stabilises under further straining until a final rise in stress and failure. During the region of constant stress a stable neck is formed which extends progressively throughout the gauge length. This is known as *cold drawing*, and during this process the molecules in the polymer are being extended in the direction of straining, leading to a highly preferred orientation in the test-piece.

The result is a necked region which is much stronger than the unnecked material, that is why the neck spreads instead of simply causing failure. When the drawing is complete, the stress–strain curve rises steeply to final fracture.

Toughening of Polymers

The moderate ductility apparent in curves 3 in Fig. 5.4 is more difficult to achieve. One approach is to prepare a two-phase material by a technique known as *copolymerisation*. An example of this is acrylonitrile-butadiene styrene which consists of spheroidal inclusions of the soft elastomeric polybutadiene in a matrix of the relatively rigid styrene acrylonitrile (SAN) copolymer. When the material is strained, stress concentrations form in the matrix around the inclusions. Defects then grow from the inclusions, absorbing extra energy as they do so – the process being known as *elastomer toughening*, although the overall tensile modulus of the material is reduced by the change in microstructure.

Another approach has been to synthesise new polymers based on different kinds of repeat units, for example the use of amorphous polycarbonate. This contains rigid aromatic groups in the molecular backbone, and exhibits mechanical toughness well below its glass-transition temperature. This may again arise by the formation of energy-absorbing defects as happens in 'elastomer toughening' described above.

Fracture

Crazing

If a brittle transparent polymer such as polystyrene (widely used for the manufacture of rulers and the bodies of ballpoint pens) is loaded in tension at room temperature, plastic deformation leads to the appearance of small, white, crack-shaped features called *crazes*, typically some 5 μm in thickness. Crazes usually develop at orientations which are at right-angles to the principal stress axis, but they are not in fact cracks but a precursor to fracture.

Crazes are a form of highly localised yielding, and often start from scratches on the specimen surface. The growth of a craze occurs by the extension of its tip into uncrazed material. They are regions of cavitated material, bridged by drawn and oriented polymer 'fibrils' of approximately 20 μm diameter, as indicated in the sketch of Fig. 5.7. The craze thickens by lengthening of the fibrils. The yielding is thus localised into these crack-shaped regions, being constrained by the surrounding undeformed solid, giving rise to the presence of

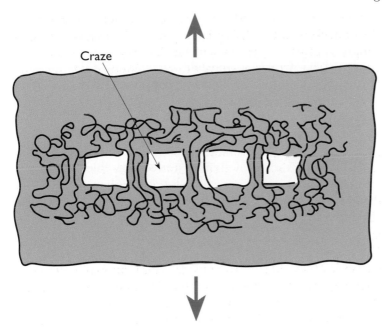

Craze

Fig. 5.7 Crazing in a linear polymer: strong strands bridge the microcrack.

local hydrostatic tensile stresses. The voids are interconnected, being typically 10–20 nm is size, lying between the oriented fibrils, so that the density of the craze is only about half of that associated with the uncrazed matrix. Crazing thus results in an overall increase in volume of a polymer which yields in this way.

The criterion for the nucleation of a craze in a three-dimensional stress state is of the form:

$$\sigma_{\max} - \sigma_{\min} \geq \Lambda(T) + \frac{\Omega_1(T)}{\sigma_H}$$

where $\sigma_{\max}$ and $\sigma_{\min}$ are maximum and minimum principal stresses, respectively, $\Lambda(T)$ and $\Omega_1(T)$ are material constants which depend on the temperature, and σ_H is the hydrostatic stress.

Crazes grow slowly at first and absorb considerable amounts of energy, although they contribute towards an overall weakening of the material. They eventually turn into true cracks which propagate rapidly.

Brittle Fracture

Brittle fracture is often observed in plastics, and in common with other materials it originates from defects which have a stress concentrating effect. Such defects can arise from the processing of the material, service damage or from poor design.

Because of their non-linear elastic response, and possibility of localised crack-tip plasticity, it may not be appropriate to apply linear elastic fracture mechanics methods to the testing of polymers. Crack-opening displacement (COD) measurement (see Chapter 2) is important in relation to plastics. Values of K_{Ic} are obtained under plane strain conditions using thick samples, and there is a problem with plastics in fabricating good quality and representative thick samples.

A further problem with polymer fracture toughness assessment is that crack growth may not be unstable, but may be time-dependent and occur where K is less than K_{Ic}. A simple model which may be applied to plastics describes the crack growth rate, da/dt in terms of material constants C and m, i.e.:

$$da/dt = C\ K_I^m$$

Impact Tests

When plastics are deformed under impact loads, the molecular structure may be unable to relax at the high applied rate of strain, and so the material may fracture in a brittle manner due to breakage of the molecular chain. Various empirical tests have been devised to measure the susceptibility of polymers to impact loads: these include machines in which the potential energy of a pendulum is used to fracture the sample. These are simple and quick to perform, and the test-piece contains a sharp notch which produces a triaxial state of stress. Many plastics show a transition from ductile to brittle behaviour when tested over a range of temperature, the exact temperature depending on the loading-rate and the notch sharpness. Such tests are thus essentially *qualitative* in character, and are of limited value in their ability to predict the impact resistance of particular products.

Product impact tests are often written into standards for polymer products. Thus samples of PVC pipes or other extruded products may be subjected to a falling-weight type of test in order to assess their energy of fracture.

A **deformation map** for PMMA is shown in Fig. 5.8 which characterises the regions where particular deformation mechanisms are dominant. The diagram is thought to typify the behaviour of linear polymers, and shows deformation regions as a function of normalised stress (w.r.t. Young's modulus) versus normalised temperature (w.r.t. the glass transition temperature). In the brittle field of Fig. 5.8 the strength is calculated by linear-elastic fracture mechanics. In the crazing field the stresses are too low to make a single crack propagate unstably, but they can cause the slow growth of the cavities within the crazes. As the temperature increases,

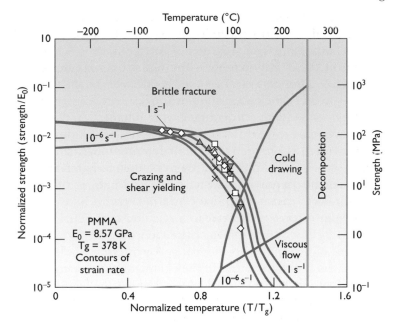

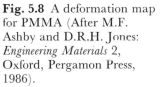

Fig. 5.8 A deformation map for PMMA (After M.F. Ashby and D.R.H. Jones: *Engineering Materials* 2, Oxford, Pergamon Press, 1986).

crazing is replaced by drawing and eventually by viscous flow. We have already discussed how the properties of polymers are strain-rate dependent, and Fig. 5.8 shows contours of constant strain-rate, so the deformation map shows how the strength varies with temperature and strain-rate.

Environment-Assisted Cracking in Polymers

Environmental stress cracking (ESC) is an important cause of embrittlement in plastics. Many polymers are strongly affected by environments such as water, vapours, or organic liquids in an analogous way to the stress-corrosion cracking of metals described in Chapter 2. A stressed plastic will fail under these conditions when it would not be expected to do so in the absence of the sensitizing environment. Again, in the absence of applied stress, the effect of the environment alone would not be deleterious, and a critical strain can be determined below which ESC does not occur. The phenomenon is thus synergistic in nature. Internal stresses resulting from moulding or welding processes may be sufficient to cause problems.

The most serious example of ESC is the oxidative cracking of rubber. The 'perishing' of both natural and synthetic rubbers arises from interaction with the ordinary atmosphere, often accelerated by ultraviolet radiation. The presence of carbon black or other special additives alleviates the effect, since they preferentially absorb the harmful radiation and convert it to thermal energy.

ESC originates from microscopic surface defects where the active medium interacts with the high stress region at the crack tip. The material becomes locally weakened and the crack spreads. The commonest examples are those involving amorphous polymers in contact with organic solvents (e.g. polycarbonate in contact with low molecular weight hydrocarbons such as acetone), although detergents can cause ESC failure when in contact with semi-crystalline materials such as polythene. In polymers which craze under stress, the liquid environment enters the crazes and penetrates between the molecular chains, giving rise to a swelling of the structure.

This local swelling effect of the environment reduces the amount of external work necessary for fracture to about 0.1 Jm^{-2}, which is of the order of the chemical bond energy, and several orders of magnitude lower than that required to fracture most tough plastics in the absence of ESC. The rate-controlling mechanism for the cracking process appears to be the diffusion of the deleterious medium in the polymer structure, the process being assisted by the applied stress. Resistance to ESC can usually be enhanced by using material of higher molecular mass, so that longer molecular chains bridge the craze and inhibit the growth of a crack within it.

The propensity for ESC is often assessed by empirical standard tests. One simple test involves immersing bent strips of plastic in a particular medium, and subsequently to examine them for signs of defects. Among the parameters that can be measured is the time to initiate visible damage such as crazing.

Fatigue

When subjected to fluctuating strains, polymers may fracture by fatigue, although , in contrast to the behaviour of metals, there are two processes by which the failure may occur. One process is analogous to that encountered in metallic materials, in that fatigue cracks may initiate and propagate to final failure. The other process arises from the hysteretic energy generated during each loading cycle. Since this energy is dissipated in the form of heat, a temperature rise will take place when isothermal conditions are not met. This temperature rise can lead to melting of the polymer, and failure of the component occurs essentially by viscous flow.

Thermal Fatigue

Consider a polymer subjected to a sinusoidal variation of

cyclic stress, $\sigma = \sigma_0 \sin \omega t$, where σ is the stress at time t, σ_0 is the peak stress and $\omega = 2\pi f$, f being the test frequency. The viscoelastic response of the polymer implies that the corresponding variation in strain will be out of phase with the stress by a phase angle δ. The peak values of the stress and strain, σ_0 and ϵ_0, are related by the complex modulus E^*, where $E^* = E' + iE''$. E' is known as the storage modulus and E'' as the loss modulus.

One can also define a complex compliance, $D^* = 1/E^* = D' - iD''$, where D' and D'' represent the storage compliance and loss compliance respectively.

From a consideration of the energy dissipated in a given cycle, and by neglecting heat losses to the surrounding environment, the temperature rise per unit time (ΔT) may be described by the equation:

$$\Delta T = \frac{\pi f D''(f, T)\sigma_0^2}{\rho c_p}$$

where ρ is the density and c_p the specific heat of the polymer. As the temperature of the specimen rises, softness results in larger deflections for a given applied stress amplitude. These larger deflections will in turn give even greater hysteretic energy losses until a point is reached when the specimen can no longer support the applied load. $\Delta\sigma - N_f$ curves can be constructed (Fig. 5.9) which show three distinct regions, marked I, II and III. Region III represents an endurance limit below which heat is dissipated to the surroundings as rapidly as it is generated, and the temperature of the specimen stabilizes at a value which is insufficient to cause thermal failure after 10^7 cycles.

Thermal fatigue may be reduced by decreasing the test frequency, cooling the test sample, or increasing the surface-to-volume ratio of the testpiece. The introduction of intermittent rest periods allows the specimen to cool, and is another means of achieving an increased fatigue life.

Mechanical Fatigue

The existence of Region I (Fig. 5.9) depends on whether crazes form at high values of $\Delta\sigma$, and whether the crazes cause microscopic cracks to nucleate. Polystyrene (PS) and polymethylmethacrylate (PMMA) are both glassy polymers that can readily form crazes, and both materials exhibit a well-defined region I. At the higher $\Delta\sigma$ end of Region II, *slow growth* of crazes and their transformation into cracks are dominant mechanisms of failure. In contrast, the life in

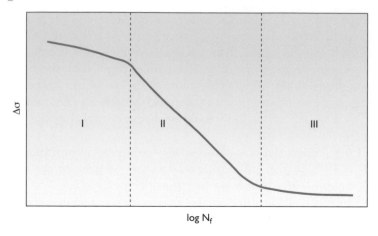

Fig. 5.9 Schematic *S–N* curve for polymers.

region III is controlled by the incubation time for the *nucleation* of microscopic flaws.

Fatigue Crack Growth (FCG)

The FCG behaviour of a wide variety of amorphous and semicrystalline polymers can be characterised in terms of the stress intensity factor, ΔK, Fig. 5.10(a) for PMMA, epoxy, polycarbonate (PC), Nylon 66, polyacetal (PA), and poly (vinylidene fluoride) PVDF (After Hertzberg, Nordberg and Manson, 1970 and Hertzberg, Skibo and Manson, 1979). Since these materials have very different elastic moduli, for

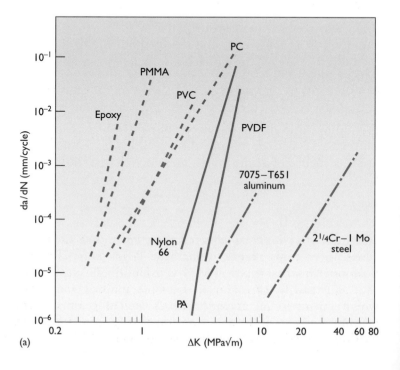

Fig. 5.10a FCG characteristics of some amorphous (dashed lines) and semi-crystalline (solid lines) polymers. Comparative data for a steel and an aluminium alloy are shown.

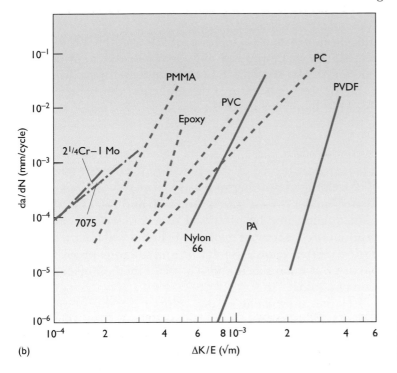

Fig. 5.10b FCG rates shown in Fig. 5.10a replotted as a function of $\Delta K/E$.

comparison purposes it is instructive to plot da/dN against $\Delta K/E$, where E is the relevant Young's modulus. An example is shown in Fig. 5.10 (b), (after Hertzberg, Skibo and Manson, 1979, and Manson, Hertzberg and Bretz, 1981) and in both diagrams data for two engineering alloys are included for comparison, namely a $2\frac{1}{4}$Cr–1Mo steel and a 7075 aluminium alloy, and it is evident that plastics will exhibit superior or inferior FCG resistance compared with metals depending on whether cycling is conducted under strain-control or stress-control, respectively.

Among the polymers, it appears that in general semicrystalline polymers exhibit a superior FCG resistance to amorphous ones.

THE JOINING OF POLYMERS

The question of joining similar or dissimilar polymers often arises, as it is frequently necessary to assemble two subcomponents when a product is manufactured. Again, items are commonly fabricated from sheet or semifinished sections, which requires the employment of fastening or welding techniques. We will consider three modes of joining polymers, welding, adhesives and mechanical fasteners.

The Welding of Polymers

Since welding involves melting and flow of the surfaces to be joined, the technique will only be applicable to *thermoplastic* polymers. The workpiece may have its temperature raised either by the application of thermal energy or by performing local mechanical work of friction.

Friction Welding

This can be achieved by several techniques; the simplest is the spin welding of two thermoplastics at relative speeds of up to 20 ms^{-1} under pressures of between 80 and 150 kPa. Welds of high quality may be produced in a few seconds, although residual stresses may be generated. Tubes and hollow sections can be welded very satisfactorily, and since the process can be carried out in liquids it is also a useful method of encapsulation of liquids.

Relative movement of the components by vibration in linear oscillation may also be employed. This method of friction welding is widely used in the automotive manufacturing industry to produce large, complex joints. A development of this principle is *ultrasonic welding*, in which the parts to be joined are held together under pressure while mechanical vibrations perpendicular to the area of contact are applied by means of a piezo-electric transducer at frequencies in the range 20–40 kHz. As the energy output of these devices is limited, the size of possible weld is much smaller than that in normal vibration welding, and tooling is expensive, but the method is well-suited to mass production and finds wide use in industry in the assembly of domestic products. No heat is required, and joint strengths approaching 100% of that of the parent materials are readily achieved.

External Heating Methods

Hot tool welding employs an electrically heated flat plate which is sandwiched between the two pieces to be joined. When a temperature of 180–230°C (depending on the particular polymer) has been achieved, the plate is withdrawn and the surfaces are pressed together under a specified stress for sufficient time for a joint to be made. In the case of items of large cross-section, such as large pipes, this time can be quite protracted – several tens of minutes is common – but very strong joints can be produced, with strengths at least 90% of the parent material. It is essential that the surfaces to be joined are clean, for a successful weld to be achieved.

A variant of this approach is the use of an 'electrofusion connector', for joining plastic piping. It consists of a coiled electric heating element embedded near the inside surface of a specially constructed joint made of the same plastic as the pipes to be joined. The joint is assembled, a current passed, and the joint fuses with the pipe material. Rapid weld times can be achieved, although the joint strength may be impaired by the presence of the heating element.

Hot gas welding heats a filler rod and the edges of the workpiece to be welded by means of a stream of hot gas from a welding gun. Compressed air is normally employed, but if the polymer can undergo degradation by oxidation, then a nitrogen stream is used. Temperatures between 200–300°C are achieved, and the technique can be applied to a wide range of thermoplastics, if necessary in the form of large complex components.

There is a danger of entrapped air pockets when this technique is used, with a resulting risk of fracture in service if the polymer is notch-sensitive.

Adhesive Bonding

This technique was discussed in Chapter 3 in the context of the joining of metals, but their use is also widely encountered in the joining of polymers. Good wetting of the polymer by the adhesive is required, and will be achieved if there is a strong chemical bond formed between the adhesive and the adherands.

There are three important categories of adhesives which may be used for the joining of polymers:

1. Hot melt adhesives.
These are thermoplastics such as PE or PET which are melted and applied to the adherands which are then squeezed together during the cooling cycle. Although good bond strengths may be achieved, the joint may creep in service if the temperature is not low.

2. Solvent-based adhesives.
Amorphous plastics are the most likely to dissolve in appropriate solvents, and these include materials such as PS, ABS, PVC and PC. Solvent alone will form an adhesive joint, but polymer solutions have better gap-filling properties, and are available in various viscosities depending on the application. Many of these adhesives are based on rubber, and are used as 'contact' adhesives to form tough joints of fairly low strength.

3. Reaction cured adhesives.

Very high bond strength are achievable with this class of adhesive: polymerization and cross-linking takes place after mixing low-viscosity precursors. Such adhesives are usually temperature and solvent resistant, the main types being:

(i) epoxies
(ii) Phenolic
(iii) cyanacrylates, widely used in the bonding of rubber, and
(iv) anaerobics, which cure when air is excluded (useful in thread-locking applications).

Mechanical Fastening

Snap Fitments

Designs of this nature are possible using semi-crystalline thermoplastics (e.g. PP, PE, Nylons) which exhibit resilience and high elastic strains.

Screw Fitments

Self-tapping screws are the most commonly used form of mechanical fastener for polymers. Thermoplastics can employ *thread-forming screws*, where elastic relaxation processes ensure a tight fit. Thermosets are too brittle for this technique, and they tend to crack in use, so recourse has to be made to *thread-cutting screws* for these materials.

POLYMER DEGRADATION

Plastics are frequently preferred to metals for use in structural applications because of their resistance to corrosion. They are often regarded as more corrosion resistant than metals, but this view is an oversimplification of the situation and there are in fact a number of ways in which polymeric materials may degrade over a period of time. These are as follows:

1. Oxidative degradation
2. Radiation degradation
3. Mechanical degradation
4. Microbiological degradation, and
5. Chemical attack.

We will consider these in turn.

Oxidative degradation is an autocatalytic process of attack on the hydrogen atoms, to form hydroperoxides. The stability of polymers is thus inversely proportional to the number of hydrogen atoms with the carbon atoms present in the polymer chain. The degradation is catalyzed by heavy metals such as copper.

The degradative reaction can be inhibited by the presence of hydrogen-donating compounds such as hindered phenols or by peroxide decomposers. Mixtures of different types of stabilisers appear to be synergistic in their effect. Natural rubbers and other elastomers can be protected against attack by ozone by the addition of microcrystalline wax. More permanent stabilisation is obtained by the addition of derivatives of the phenol *p*-phenylenediamene.

Radiation degradation. Engineering plastics are commonly used out of doors, and it is now recognised that sunlight with a wavelength less than 290 nm is responsible for the photooxidation of polymer surfaces. Long wavelengths have insufficient energy to break covalent bonds in organic compounds, but ultraviolet wavelengths will selectively excite electrons in the polymer chain. This increases the vibrational and rotational energy of the covalent bonds, leading to degradation by bond cleavage. Fortunately, the intensity of light in the 290–320 nm wavelength region is estimated to account for no more than 0.5% of the radiant energy of the sunlight at noon in southern regions. If this were not the case, few plastics would be of use outside.

Typical types of degradation include yellowing, chalking, surface embrittlement, loss of tensile or impact strength, and cracking. The process usually proceeds from the surface layers, and the weakened surface can act as a site for crack nucleation. Such cracks may then propagate into the undegraded material beneath, causing failure.

No photochemical changes occur when the light is dissipated harmlessly as heat, and pigments such as carbon black are commonly added to absorb ultraviolet radiation and then re-emit it as thermal energy. Hydroxybenzophenones and benzotriazoles are also widely used as ultraviolet absorbers.

Aliphatic polymers such as PE are degraded by gamma radiation, although aromatic polymers, such as PS, are relatively resistant to high-energy radiation. PP (polypropylene) utensils may be embrittled by radiation damage if they are subject to sterilization treatment by doses of gamma radiation of 2.5 Mrad. Equipment should be designed to last 40 years when subjected to a dose of 20 Mrad.

Mechanical degradation may occur when stress is imposed on a polymer through machining, stretching or ultrasonics. Bond cleavage may occur, forming macroradicals which can add oxygen and produce compounds which will undergo degradative reactions.

Microbiological degradation. Most widely used plastic materials are inert in the presence of microbes, and this stability is important in many applications. Only short-term performance is required in certain situations before the material is discarded – for example in fast food packaging. It is considered an advantage if the discarded plastic degrades when exposed to microbes, and it is a challenge to polymer scientists to develop plastics that possess the requisite properties for their anticipated service life, but which are eventually capable of degrading in a safe manner. In 1987, 6×10^9 kg of plastics were used for short-term packaging applications, and this figure will exceed 60×10^9 kg by the turn of the millennium.

We may define biodegradable plastics as those whose physical integrity is lost upon contact with microbial and/or invertebrate activity in a natural environment within a limited period of time. In the limit, this represents conversion of the material to carbon dioxide, water, inorganic salts, microbial cellular components and miscellaneous by-products.

Most plastics at present used for packaging consist of high- and low-density PE, which do not degrade by microbiological action. Due to their large chain lengths and high molecular weight, most widely used alkane-derived plastics may have lifetimes of hundreds of years when buried in typical solid-waste disposal sites. Low molecular weight hydrocarbons can be degraded by microbes, but the rate of degradation becomes very slow when the length of the polymer chain exceeds 24 to 30 carbon atoms. Decreasing typical polymer molecules to biologically acceptable dimensions requires extensive destruction of the PE matrix. This destruction can be partly accomplished in blends of PE and biodegradable natural polymers by the action of macroorganisms such as arthropods, millipedes, slugs, snails, etc.

Starch-polyethylene complexes have been manufactured which exhibit physical properties approaching those of low density PE. The starch is present as a separate phase and is attacked by fungi and bacteria. This weakens the polymer matrix, and the eventual breakdown of the polymer chains reduces the molecular weight which enhances microbial attack.

Chemical attack. Plastics are susceptible to environmental failure when exposed to certain organic chemicals, and this limits their use in many applications. Even aqueous media can cause degradation, however, although the processes involved differ from the corrosive attack of metals in such environments.

Diffusion of species *into* plastics is common, and adverse effects can arise which are not chemical in nature. In most interactions of water with structural plastics no chemical bonds are altered, but damage known as 'physical corrosion' may occur. Absorbed moisture has been shown to act as a plasticizer, reducing the glass transition temperature and the strength of the polymer. These effects are essentially reversible, although other, irreversible, effects may be encountered, such as microcracking or crazing, as well as chemical degradation of the polymer structure.

Organic liquids, such as cleaning fluids, detergents, petrol and lubricants may seriously reduce the mechanical properties of plastics. As already discussed, the most serious problems arise when a material is exposed to aggressive fluids when it is under stress. Organic liquids may interact both chemically and physically with a polymer. Chemical interactions may involve a decrease in the molecular weight by chain breakage; this in turn may cause a reduction in mechanical properties such as tensile strength, stiffness and fracture toughness. Figure 5.11 illustrates the decrease in tensile strength of polyester-base polyurethane (PUR) as a function of exposure time in methanol. The methanol is believed to swell the PUR, and also causes molecular weight reduction through chain breakage.

As an example of physical interaction when a plastic component is exposed to an organic chemical, aggressive molecules may diffuse into the material leading to plasticization. Swelling of the polymer results in high stresses which can cause crazing or cracking. Fracture arising from physical effects has been observed in many glassy plastics such as PMMA, PS and PC because of anisotropic swelling.

MODELLING OF POLYMER STRUCTURE AND PROPERTIES

Advances in computer technology have stimulated the construction of models describing the microscopic structure of polymeric materials and the use of such models to predict

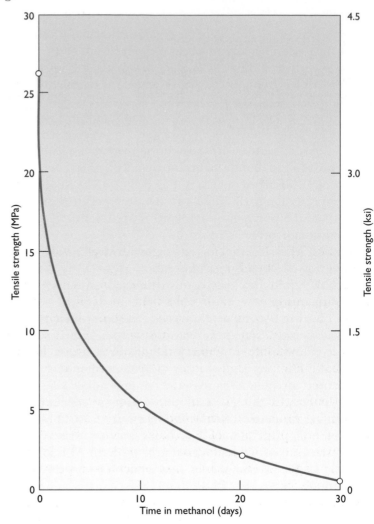

Fig. 5.11 Tensile strength of PUR aged in methanol at 60°C as a function of exposure time.

polymer properties. This type of modelling attempts to establish quantitative relationships between processes at the molecular level and the behaviour at the macroscopic level. This approach can be used as a design tool when it has been successfully tested against experimental data, permitting the identification of the chemical constitution, and the synthesis or processing conditions required to yield a material with certain prescribed characteristics.

Given a set of performance requirements, it is necessary to identify the monomers to be used, the process of synthesis to be adopted, and the processing conditions to which the material has to be subjected in order to meet the specified

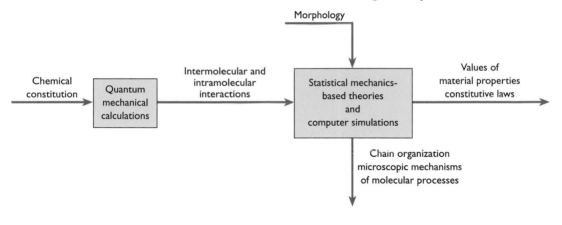

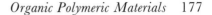

properties. The first two of these stages fall largely within the fields of organic chemistry and chemical engineering, and the third will include finite-element calculations of melt flow in processing equipment or of stress fields in deformed structures. The final stage is to describe relationships between chemical constitution, polymer morphology (e.g. the type and size of crystallites in semicrystalline polymers, or the size and distribution of domains in others), and macroscopic properties.

Molecular modelling is employed to provide this final link, by deriving values of properties or constitutive laws based on fundamental principles of molecular science. Figure 5.12 illustrates schematically the basis of molecular modelling approaches to the establishment of structure-property relations in polymers.

Quantum mechanics can be used to derive the bond lengths, bond angles, and intermolecular interactions in polymers, while statistical mechanics provides a link between macroscopically observable behaviour and molecular geometry and energetics, elucidating molecular mechanisms that govern material properties. Commercial software packages are available which incorporate theoretical and simulation techniques, so that modelling methods (in conjunction with experimental research) are now employed for the development of new polymeric products in industry.

Atomistic simulation methods are successful in predicting equation-of-state behaviour of polymer melts and glasses of moderate molecular weight, as well as the elastic constants of polymer crystals and glasses. Molecular modelling is thus clearly emerging as an important tool in polymer science and engineering.

Fig. 5.12 Information flow involved in molecular modelling for the establishment of structure-property relationships.

READING LIST

F.W. Billmeyer Jr: *Textbook of Polymer Science*, Wiley, New York, 1984.

D.W. Clegg and A.A. Collyer: *The Structure and Properties of Polymeric Materials*, The Institute of Materials, London, 1993.

L.R.G. Treloar: *The Physics of Rubber Elasticity*, Oxford University Press, 1975.

M.F. Ashby and D.R.H. Jones: *Engineering Materials 2*, Pergamon, 1986.

Computer simulation of polymers, R.J. Roe ed., Prentice-Hall, Englewood Cliffs, NJ, USA, 1991.

General Reference

Engineered Materials Handbook, *Volume 2*, *ENGINEERING PLASTICS*, ASM International, Metals Park, OH USA, 1988.

6 Composite Materials

INTRODUCTION

A composite consists of a matrix material, dispersed within which is a dispersion of one or more phases of another material. In successful composites, the product has a combination of properties which is superior to those of the individual components. Although the dispersed phase can be in the form of particles, in many important examples it is in the form of fibres, and *fibre composite* materials form an important subset of this class of engineering materials. Natural composite materials such as wood and bone have been used by mankind for many thousands of years: these are based on fibres of cellulose and collagen respectively. The advantages of deliberately combining materials to obtain improved properties have been recognised since biblical times – the fifth chapter of the book of Exodus refers to the importance of the incorporation of straw in the manufacture of bricks – and many similar examples may be quoted throughout the history of civilisation.

There are many reasons for making composites: the incorporation of fibres into brittle ceramics referred to above produces a composite of enhanced toughness. Fillers, such as the presence of aggregate in concrete, reduce the overall cost of the product, and additionally improve the compressive strength. The second phase may furthermore be a gas, as in the manufacture of foamed products of low density.

On the basis of strength and stiffness alone, fibre reinforced composite materials may not be superior to metals of comparable strength, but when the specific modulus (i.e. modulus per unit weight) and specific strength are considered, then their use implies that the weight of components can be reduced. This is an important factor in all forms of transport, where reductions of weight result in greater energy savings. We will first discuss how composite materials may be fabricated, and then consider the extent to which their mechanical properties may be understood in terms of some simplified models.

THE MANUFACTURE OF COMPOSITE MATERIALS

As stated above, fibre composite materials represent a major part of this category: a wide range of possible fibres exists, and Table 6.1 illustrates the range of properties they possess.

The most widely used fibres are of carbon, glass and Kevlar, and the method of their production is outlined below.

Carbon fibres consist of small crystallites of graphite, whose crystal structure is shown in Fig. 1.2(b). The atoms in the basal planes are held together by very strong covalent bonds, and there are weak van der Waals forces between the layers. To obtain high modulus and high strength the layer planes of the graphite have to be aligned parallel to the axis of the fibre, and the modulus of carbon fibres depends on the degree of perfection of alignment of the atom planes. This varies considerably with the particular manufacturing route adopted, of which there are three main possibilities:

(a) Starting with the polymer PAN (polyacrylonitrile),

Table 6.1 Mechanical properties of some reinforcing fibres.

Material	Density ρ (kg m^{-3} × 10^3)	Young's Modulus E (GPa)	Tensile Strength $\sigma*$ (GPa)	Fibre Radius r (μm)
E-glass fibres	2.56	76	1.4 – 2.5	10
Carbon fibres (high modulus)	1.75	390	2.2	8.0
Carbon fibres (High strength)	1.95	250	2.7	8.0
Kevlar fibres	1.45	125	3.2	12
Silicon carbide (Mono-filament)	3.00	410	8.6	140
Silicon carbide (Nicalon)	2.50	180	5.9	14
Alumina (Saffil)	2.80	100	1.0	3

which closely resembles polyethylene in molecular conformation, it is converted into a fibre and then stretched to produce alignment of the molecular chains along the fibre axis. While still under tension, it is heated in oxygen to form cross-links between the ladder molecules and finally chemically reduced to give (at high temperatures) a graphitic structure. The final graphitisation temperature determines whether the fibres have maximum stiffness but a relatively low strength (Type I fibres), or whether they develop maximum strength (Type II).

(b) Alternatively, fibres may be produced by melt-spinning molten pitch. During the spinning process, the orifice causes the planar molecules to become aligned. It is then treated, whilst held under tension in order to maintain its preferred orientation, at temperatures up to $2000°C$ to form the requisite grains of graphite.

(c) It is also possible to stretch either of the fibre types described above during the graphitization stage, giving further orientation of the layers parallel to the fibre axis.

Glass fibres are commonly produced in 'E-glass' (E is for electrical), because it draws well and has good strength and stiffness. A typical composition (wt %) would be 52 SiO_2, 17 CaO, 14 Al_2O_3, 10 Ba_2O_3 with some oxides of Mg, Na and K, and molten glass is gravity-fed into a series of platinum bushings each of which has several hundred holes in its base. Fine glass filaments are drawn mechanically as the glass exudes from the holes, then wound on to drums at speeds of several thousand metres per minute.

The strength of the glass fibres is dependent upon the surface damage arising when they rub against each other during processing. The application of a size coating early at an early stage during manufacture minimises this degradation in properties, by reducing the propensity for forming these 'Griffith' cracks. The size consists of an emulsified polymer in water, and also has the effect of binding the fibres together for ease of further processing.

Organic fibres of high strength and stiffness may be manufactured, and one of the most successful commercial organic fibres has been developed by the Du Pont company with the trade name of **Kevlar**. It is an aromatic polyamide, and the aromatic rings result in the polymer molecules having the properties of a fairly rigid chain. The fibres are produced by extrusion and spinning processes which align the stiff polymer molecules parallel to the fibre axis.

We will now consider how various types of composite materials may be fabricated.

Polymer Matrix Composites

Fillers For Plastics

Apart from fibres, a wide range of other solids may be added to plastics in order to modify their properties or to reduce their cost. These additives are known as *fillers*, and many of them are based on ground minerals such as limestone. A finely ground mineral may be mixed with a polymer such as polypropylene, which is then formed and moulded by normal methods. Other minerals used include silicates such as kaolin, talc and quartz. Glass is also used as a filler in the form of solid spheres, flakes, or hollow microspheres. These mineral fillers will contribute to an increase in stiffness of the polymer (see below).

Natural products such as wood flour can be used to provide low-cost fillers, and there are certain special fillers that are introduced to confer a fire-retardance to the material. Alumina trihydrate is in the latter category, since above $200°C$ it decomposes to produce alumina and about 35% by weight of water.

Fibre Reinforced Plastics

Either thermosetting resins or thermoplastics may be used as matrices, and their selection is determined by their ease of fabrication for a particular product.

There are numerous manufacturing routes, which we can group into two types of process, namely *closed mould processes* and *open mould processes*.

Examples of closed mould processes include:

(a) *Injection moulding* in which molten polymer, mixed with short fibres, is injected into a split mould, where it solidifies or cures,
(b) *Resin injection* into the mould cavity which contains fibres in cloth form, and
(c) *Cold press moulding*, where the fibres are impregnated with resin before pressing between matched dies.

Open mould processes include:

(a) *Filament winding*, where the fibres are fed through a bath of resin before winding on to a mandrel. This forms a tubular component, which is removed from the mandrel after the resin is cured, and
(b) *Hand lay-up,* in which fabrics made from the fibres are placed on the mould and impregnated with resin, building up layers until the required thickness is achieved. The component cures without the application of heat or pressure.

(c) *Autoclave curing of pre-pregs.* Unidirectional sheets of fibres may be pre-impregnated with resin and partially cured to form a 'pre-preg'. Pre-preg sheets may be stacked on the mould surfaces in the desired sequence of orientations, and final curing is completed in an autoclave, with the assembly being consolidated by means of pressure applied from an inflated flexible bag which is placed upon the sheets.

'Smart' Fibre Composites

The idea of embedding sensors into composites during the manufacturing process dates from the 1980s. Research activity in this area has increased significantly in recent years, with several aims.

(i) By embedding sensors which can be integrated during the cure process, it will be possible to improve significantly the manufacture of advanced composites by measuring parameters such as strain, pressure and temperature.

(ii) When the component is in service, incorporated sensors would be able to monitor fatigue cracking, corrosion, overload, etc.

The development of smart composites is likely to accelerate the application of advanced fibre composite materials, for it is extremely difficult to incorporate the same capabilities in competitive materials.

Metal Matrix Composites (MMCs)

The matrix in this class of material is usually an alloy, rather than a pure metal, and there are three types of such composites, namely,

(i) *Dispersion-strengthened*, in which the matrix contains a uniform dispersion of very fine particles with diameters in the range 10–100 nm,

(ii) *Particle-reinforced*, in which particles of sizes greater than 1 μm are present, and

(iii) *Fibre-reinforced*, where the fibres may be continuous throughout the length of the component, or less than a micrometre in length, and present at almost any volume fraction, from, say, 5 to 75%.

The Production of MMCs

These can be classified into two broad categories, those in which the metallic matrix is introduced in a solid, particulate form, and those in which the metal is melted.

1. Conventional *powder metallurgical* techniques are important, in which the individual phases are mixed together in particulate form. After homogenization of the mix, the blended powders are pressed in an appropriate mould to form a 'green compact' of high porosity and low strength, and finally sintered at high temperature (in a protective atmosphere), often under external pressure ('hot pressing'), to form the final, dense, composite. An example of this type of product are high-speed cutting tools and mining drills composed of particles of tungsten carbide, WC, in a matrix of 6–20% cobalt. The WC particles are angular and equiaxed in shape, and their size is usually in the range 1–10 μm.

Hot isostatic pressing (HIP) of metal powder/fibre composites is also employed, with the materials contained in evacuated cans. Final densification may be by hot extrusion.

When particles of sub-micron dimensions are mixed with metal powders by normal powder metallurgical techniques, however, they tend to clump together, with a corresponding loss in properties. Uniform dispersions of fine particles are more readily achieved by the process of *mechanical alloying* (MA). In this technique, elemental or alloyed metal powders, together with the dispersoid are charged into a dry, high-energy, high speed ball mill. During processing, the powder particles are repeatedly fractured and rewelded to the ball surfaces until all the constituents are finely divided and uniformly distributed through the interior of each granule of powder. The MA powder is finally consolidated by extrusion under carefully controlled conditions of temperature and strain rate.

2. Liquid metal techniques.
There are a number of these techniques, including *Squeeze casting* (Fig. 6.1) whereby molten metal is introduced to a die containing a fibre preform and then pressed to cause infiltration, *stircasting*, in which the fibres or particulates are mixed with the molten metal, followed by die-casting, and *spray deposition*, in which the metal in the form of wire is fed into an atomising arc and the resulting metal vapour is vacuum sprayed on to a drum with a fibre mat wound upon it.

The disadvantage of MMCs produced by liquid metal techniques is that they offer increased performance but at increased cost, and this has led to the recent withdrawal of some major UK companies from the production of MMCs. These materials have good prospects for the future, however, and it is expected that they will experience steady incremental growth in their application in performance-limited markets.

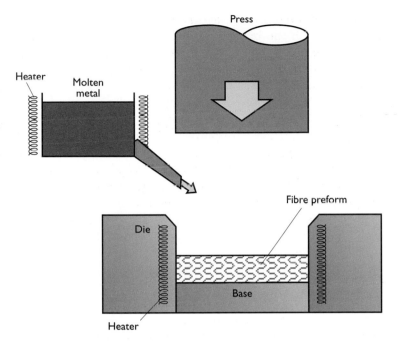

Fig. 6.1 Squeeze casting of a metal matrix composite.

Continuous fibre MMCs have found some application in aerospace structures, and automobile connecting rods have been manufactured from aluminium reinforced with fibres of stainless steel. Discontinuous MMCs with dispersions of silicon carbide fibres or particles can be shaped by standard metallurgical techniques, and these have found application in aluminium-based automobile engine components, for example in the selective reinforcement of the ring area of diesel engine pistons. This reduces weight, and increases wear resistance and thermal conductivity.

The addition of aluminium oxide fibres (Saffil) to aluminum increases the wear resistance and stiffness, while reducing the coefficient of thermal expansion, and this has also led to new uses of such MMCs in piston crowns of internal combustion engines.

CELLULAR SOLIDS

In this category we will consider cellular solids that occur in nature, such as wood and bone, as well as man-made cellular polymers such as foams. Foams are made by the entrapment of gas (either physically or chemically introduced) while the polymer is liquid, and the resultant light-weight products are used in thermal insulation, for providing buoyancy, and also in load-bearing applications such as cushioning and padding.

Wood is the most widely used of all structural materials,

Table 6.2 Mechanical properties of some woods, parallel (∥) and perpendicular (⊥) to the grain

Species	ρ-kgm^{-3}	E(∥)-GPa	E(⊥)-GPa	σ_t (∥)-MPa	σ_c (∥)-MPa	K_c (∥)-MPa m$^{\frac{1}{2}}$	K_c (⊥)-MPa m$^{\frac{1}{2}}$
Balsa	200	6.3	0.2	23	12	0.05	1.2
Mahogany	440	10.2	0.8	90	46	0.25	6.3
Ash	670	15.8	1.2	116	53	0.61	9.0
Oak	690	13.6	1.0	97	52	0.51	4.0
Beech	750	13.7	1.7	100	45	0.95	8.9
Douglas fir	590	16.4	1.1	120	50	0.34	6.2
Scots pine	550	16.3	0.8	90	47	0.35	6.1
3-plywood	520	(isotropic) 12.1					
Chipboard	720	1.9					
Hardboard	1030	4.6					

since a ten times greater volume of wood is used annually than of iron and steel. There are tens of thousands of species of trees, each of which will possess its own mechanical and structural characteristics, and furthermore there is a large variability in the properties of a particular type of tree. Typical mechanical properties of a few woods are given in Table 6.2, but from a design point of view it should be borne in mind that such properties may vary by up to $\pm 20\%$.

Wood consists of cellulose, hemi-cellulose and lignin. Lignin is an amorphous polymer, and acts mainly as a matrix for the other constituents. Cellulose consists of long-chain molecules, present in slender strands called micro-fibrils. Hemi-cellulose is similar to cellulose, and is partly amorphous and partly oriented.

Commercially important trees come from botanical classes known as 'softwoods', which are from coniferous trees, and 'hardwoods' which are from deciduous broad-leaved trees. The terms are misleading, since many softwoods are harder than many hardwoods, but the two types differ in their macrostructures, Fig. 6.2a and b.

Softwoods (Fig. 6.2a) are composed of about 90% long, slender rectangular cells called tracheids, about 3 to 5 mm in length. These can be modelled as a bundle of tubes or drinking straws, and the stiffness is much greater parallel to the axis of the tubes than in the perpendicular direction. This partially accounts for why wood is 10 to 20 times stiffer parallel to the grain than perpendicular to it. About 10% of

 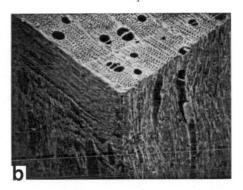

wood cells grow as rays running in the radial direction. These
ray cells act as reinforcement in this direction, increasing the
radial strength and stiffness, and they also cause the tracheids
to align themselves fairly regularly in radial rows (Fig. 6.2a).
The principal reason for the high longitudinal strength and
stiffness is found in the structure of the walls of the longitudinal cells, shown diagrammatically in Fig. 6.3.

Fig. 6.2 SEM images of
wood with tangential surface
on RHS and radial surface
on LHS. (a) softwood,
showing part of four growth
rings, and (b) hardwood
showing longitudinal pores.

The outer layer is mainly of lignin and hemicellulose, and
acts as a bond layer to adjacent cells. The three secondary
inner layers, S-1, S-2 and S-3 differ mainly in the helical
orientation of the microfibrils. The thick S-2 layer largely
determines the strength properties of the wood: in it the
cellulose-based microfibrils are oriented in a helix nearly
parallel to the cell's axis. Helices in the S-1 and S-3 layers
are tighter than in S-2.

Hardwoods (Fig. 6.2b) have many cell types, chiefly longitudinal fibres analogous to tracheids in the softwood, though
smaller, longitudinal vessels or pores, much larger than tracheids, and radial ray cells, which can constitute up to 30% in
hardwoods.

Wood composites. Plywood consists of a series of wood laminae
bonded together with a thermoset polymer adhesive so that
the grain direction runs at right-angles in successive layers. For
indoor use, urea formaldehyde (UF) is employed, whereas
marine ply uses phenol formaldehyde (PF) as the adhesive.

Chipboard is formed by the compression moulding of
wood chips bonded with about 10% UF. The material is
significantly cheaper than plywood, but its properties are
somewhat inferior.

Hardboard is made from wood chips which have been separated under pressurised steam into fibres. A mat of these fibres
is then hot pressed with UF into a board.

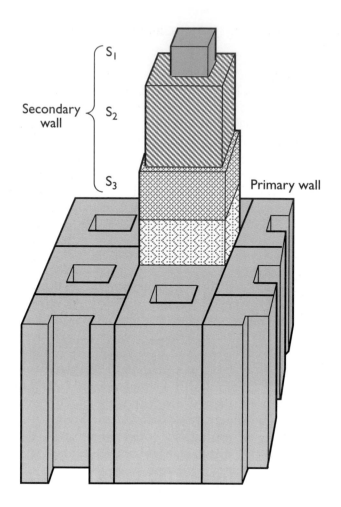

Secondary wall { S₁, S₂, S₃

Primary wall

Fig. 6.3 Idealised drawing of cell wall.

MODELLING COMPOSITE BEHAVIOUR

A number of composite models have been developed with the aim of predicting the mechanical properties of composites from a knowledge of those of the constituent phases (matrix and reinforcement). We will consider some of the simpler models below.

Stiffness

The elastic behaviour of composite materials can be modelled most readily if simplifying assumptions are made. One such simplification is to assume that the two components have identical Poisson's ratios. It is also commonly assumed *either* that the elastic strain is uniform throughout the composite, which implies differences in stress distribution, *or* that the elastic stress field is uniform, implying variations in local strain.

(a) Assuming uniform strain within the composite.
Consider a composite material consisting of two phases, 1 and 2, which possess a similar Poisson's ratio but differing Young's moduli of values E_1 and E_2. Let us assume that, when the material is elastically distorted, the macroscopic stress and strain are reproduced in a typical unit volume which consists of a *single particle* of material 2 in a cube of matrix material (1). Assume that the cube is loaded across two opposite faces by a force F (Fig. 6.4), and considering a cross-section of the composite of thickness dx at a distance x from an end face, which intersects an area A_1 of matrix and an area A_2 of the dispersed material, if the strain (ϵ) is uniform within this element, we can write the normal stress on area A_1 as $E_1\epsilon$, and that on area A_2 as $E_2\epsilon$.

The total force on the cross-section must equal F, so,

$$F = A_1 E_1 \epsilon + A_2 E_2 \epsilon$$

and since $A_1 + A_2 = 1$, we can write

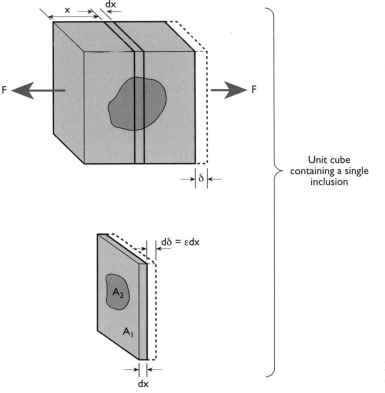

Unit cube containing a single inclusion

Fig. 6.4 Unit cube containing a single inclusion.

$$F = (1 - A_2)E_1\epsilon + A_2E_2\epsilon$$

$$= \epsilon[E_1 + (E_2 - E_1)A_2]$$

i.e. $\epsilon = \dfrac{F}{E_1 + (E_2 - E_1)A_2}$

The total elongation of the cube ($= \delta$) can be expressed:

$$\delta = \int_0^1 \epsilon(x)\,dx$$

$$= F \int_0^1 \frac{dx}{E_1 + (E_2 - E_1)\,A_{2(x)}}$$

where $A_{2(x)}$ describes the variation of A_2 as a function of x.

The value of Young's modulus for the composite, E_c, will be, for the unit cube the ratio F/δ, i.e.

$$\frac{1}{E_c} = \int_0^1 \frac{dx}{E_1 + (E_2 - E_1)A_{2(x)}} \qquad (6.1)$$

We may thus derive an expression for E_c for various volume fractions of any particular distribution of the embedded material for which $A_{2(x)}$ is a well-defined function of x. We will consider **three** examples.

Cubic Inclusions

Consider a composite material consisting of a volume fraction f of identical cubic inclusions of Young's modulus E_2 in a matrix of modulus E_1. In our unit cube of Fig. 6.4, one inclusion will be of volume f, and so the edge length of the inclusion will be $f^{\frac{1}{3}}$, and so the area of one face, A_2, will be $f^{\frac{2}{3}}$. $A_{2(x)}$ will be of the form shown in Fig. 6.5.

Substituting in equation 1.6, we obtain:

$$\frac{1}{E_c} = \int_0^{(1-f^{\frac{1}{3}})} \frac{dx}{E_1} + \int_{(1-f^{\frac{1}{3}})}^1 \frac{dx}{E_1 + (E_2 - E_1)\,f^{\frac{2}{3}}}$$

$$= \frac{1 - f^{\frac{1}{3}}}{E_1} + \frac{f^{\frac{1}{3}}}{E_1 + (E_2 - E_1)f^{\frac{2}{3}}}$$

$$= \frac{E_1 + (E_2 - E_1)f^{\frac{2}{3}} - (E_2 - E_1)f}{E_1[E_1 + (E_2 - E_1)f^{\frac{2}{3}}]}$$

So, $\dfrac{E_c}{E_1} = \dfrac{E_1 + (E_2 - E_1)f^{\frac{2}{3}}}{E_1 + (E_2 - E_1)f^{\frac{2}{3}}(1 - f^{\frac{1}{3}})} \qquad (6.2)$

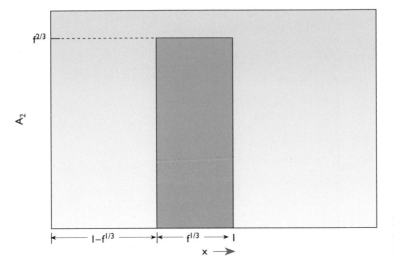

Fig. 6.5 Inclusion area vs. distance graph for cubic inclusion.

Application to Tungsten Carbide/Cobalt Cutting Tools

These so-called *cemented carbide* materials are used for high speed metal cutting tools and mining drills. They microstructure consists of tungsten carbide (WC) particles in a matrix of cobalt (Co). WC is a hard and brittle phase, and Co is a ductile metal which is able to wet the carbide and form a strong adhesive bond with it. The alloy is fabricated powder metallurgically: a fine powder of mixed WC and Co being compacted and sintered at high temperature.

The Young's modulus of cobalt is 206.9 GPa, and that of tungsten carbide is 703.4 GPa. If a series of composites of differing volume fractions of cuboidal WC particles is prepared and their Young's moduli determined, the results shown in Fig. 6.6 are obtained. The continuous line in Fig. 6.6 represents the theoretical composite moduli according to equation 6.2. It is evident that the equation well describes the stiffness of the composite.

Application to Polymers

The use of appropriate fillers of high modulus in polymers may give, in agreement with equation 6.2, a material with a modest increase in stiffness but with no increase in fabrication costs.

Continuous Fibres

If the composite consists of an array of continuous fibres parallel to the x-axis, and the moduli of the two phases are E_1 and E_2, one may again substitute in equation 6.1 but in this case the fibres will extend across the entire length of the unit

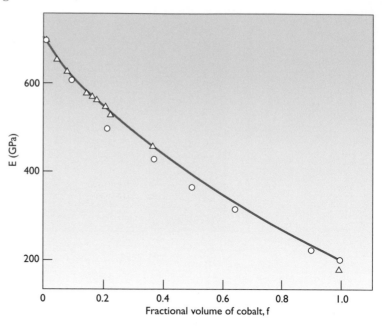

Fig. 6.6 Young's moduli of
WC–Co composites of
differing volume fraction
WC. The continuous line
corresponds to equation 6.2.

cube, so $A_{2(x)} = f_f$, the volume fraction of fibres in the material.

Substitution in equation 6.1 yields:

$$\frac{1}{E_c} = \int_0^1 \frac{dx}{E_1 + (E_2 - E_1)f_f}$$

i.e.
$$E_c = E_1 + (E_2 - E_1)\,f_f$$

$$= E_2 f_f + E_1\,(1 - f_f)$$

But $f_f + f_{\text{matrix}} = 1$, so we can write

$$E_c = E_1 f_{\text{matrix}} + E_2 f_f \qquad (6.3)$$

which predicts a linear 'law of mixtures' relationship for the modulus of the composite. This relationship has been verified experimentally with many fibre-resin systems.

A direct comparison of equations 6.2 and 6.3 has been made in predicting the moduli of composites consisting of a matrix of copper containing either continuous wires of tungsten or equiaxed particles of tungsten. The measured data are compared with the theoretical predictions in Fig. 6.7.

To give an example of greater industrial significance, Fig. 6.8 illustrates the improvements in specific stiffness achieved in aluminium-based MMCs containing either particulate SiC or aligned monofilament SiC. More than 50% improvement is readily obtained for particulate composites and over 100% for fibre reinforced systems.

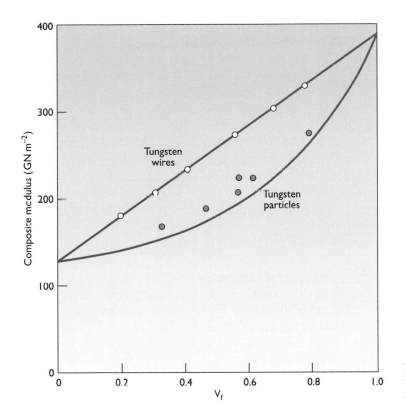

Fig. 6.7 Modulus versus volume fraction for W-Cu composites.

Continuous Lamellae

If the composite consists of on array of alternate lamellae of the two phases, then the analysis will be identical to the case for arrays of fibres, and a law of mixtures will again be predicted.

The above analyses are based on the assumption that Poissons' ratio is identical in the two phases. This is not strictly true, since different Poisson contractions will result in

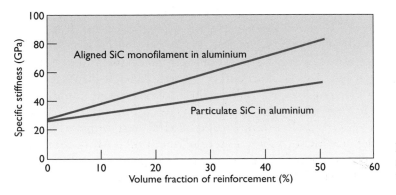

Fig. 6.8 Improvements in specific stiffness of MMCs with increasing carbide.

additional stresses which have not been considered here. The error in E_c in the direction parallel to the fibres or laminae is likely to be less than 1 or 2%, and this acceptability of the approximation is evident in the experimental data shown above.

The value predicted by the law of mixtures must be regarded as an upper bound, because the strain in the reinforcement and the matrix are not identical in practice. We have seen, however, that this approach predicts reasonably well the longitudinal properties of unidirectional continuous filament composites, and law of mixtures predictions provide upper limit goals that may be aimed at by those who design and develop composite materials.

(b) Assuming uniform stress within the composite.

Continuous Fibres or Lamellae

If a composite is loaded along an axis perpendicular to that of the fibres or lamellae (Fig. 6.9), it is reasonable to assume that the stresses in the two components are equal. In the case of fibres, which are more commonly encountered, the total strain of the composite, ϵ_c, is the weighted mean of the individual strains in the fibre (ϵ_f) and matrix (ϵ_m), i.e.

$$\epsilon_c = f_f\,\epsilon_f + (1 - f_f)\epsilon_m$$

$$= \frac{f_f\,\sigma}{E_f} + \left(\frac{1 - f_f}{E_m}\right)\sigma$$

The modulus of the composite is σ/ϵ, so that

$$E_{c\perp} = \left\{\frac{f_f}{E_f} + \frac{1 - f_f}{E_m}\right\}^{-1} \qquad (6.4)$$

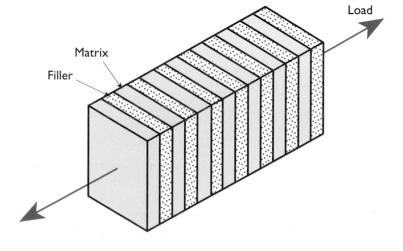

Fig. 6.9 Lamellar composite with loading axis perpendicular to the lamellae.

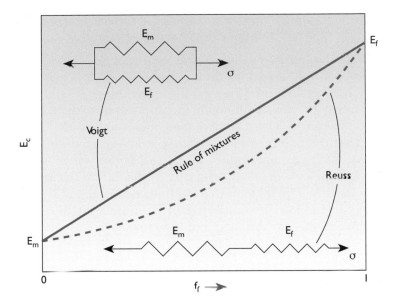

Fig. 6.10 Moduli of composites versus volume fraction, according to equations 6.3 and 6.4.

When plotted, equations 6.3 and 6.4 appear as shown in Fig. 6.10, and these values of composite moduli indicate why fibre reinforced composites are so anisotropic in their properties. For example, the modulus of wood measured parallel to the fibres is about 10 GPa, but it is only about 1 GPa perpendicular to the grain. The **degree of anisotropy** (κ) of a material may be expressed as

$$\kappa = \frac{E_{\parallel}}{E_{\perp}} - 1$$

in terms of the moduli parallel to and perpendicular to a principal axis.

Discontinuous, Non-Aligned Fibres

In a composite with discontinuous fibres, or one in which the fibres are aligned over a range of directions, equation 6.3 cannot predict the composite modulus in the nominal fibre direction. The fibre orientation distribution is difficult to assess, and the effect of such misalignment may be incorporated into the rule of mixtures (equation 6.3) by including an *efficiency factor*, B, so that the equation becomes:

$$E_c = B E_1 f_{\text{matrix}} + E_2 f_f$$

where B is unity for complete alignment, is $\frac{1}{2}$ if the fibres are aligned in two directions at right-angles, stressed in one of these directions, is $\frac{3}{8}$ if the fibres are randomly distributed in a plane and the composite is stressed in the plane, and is 0.2 if the fibres are randomly distributed in three dimensions.

The Tensile Strength of Fibre Composites (σ_c)

Aligned, Continuous Fibre Composites

We will assume that the fibres and the matrix possess a similar Poisson's ratio, and that they are well-bonded so that they deform together. The load on the composite will be shared between the two components in proportion to their cross-sectional areas. There are two possibilities to consider:

(i) If the failure strain of matrix < failure strain of fibre (Fig. 6.11). At low values of f_f, the failure stress of the composite, $\sigma_c{}^*$, depends primarily of the failure strength of the matrix, $\sigma^*{}_m$. The matrix fractures before the fibres and then all the load is transferred to the fibres: the fibres are unable to support this load and they break, so

$$\sigma_c{}^* = \sigma_f\, f_f + \sigma_m{}^*(1 - f_f) \qquad (6.5)$$

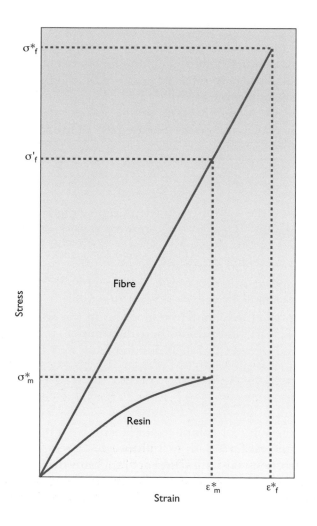

Fig. 6.11 Stress–strain curves of fibre and matrix with failure strain of matrix < that of fibre.

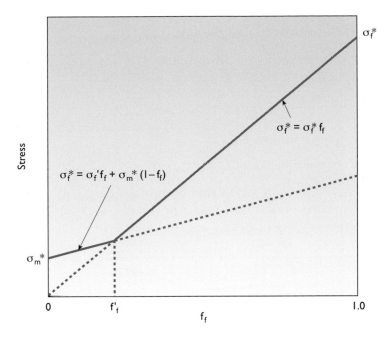

σ_f^*

$\sigma_f^* = \sigma_f^* f_f$

$\sigma_f^* = \sigma_f' f_f + \sigma_m^* (1 - f_f)$

Stress

σ_m^*

0   f'_f 1.0

f_f

Fig. 6.12 Variation of fracture stress of composite with volume fraction of fibres, when $\epsilon_m^* < \epsilon_f^*$.

When f_f is large, the matrix takes only a small proportion of the load, because $E_f > E_m$, so that when the matrix fractures the fibre do not experience a sufficient increase in load to cause them to fracture and the load on the composite can be increased until the fracture stress of the fibres (σ_f^*) is reached, so

$$\sigma_c^* = \sigma_f^* \cdot f_f \tag{6.6}$$

The fracture strength of the composite varies with f_f as illustrated in Fig. 6.12, and this analysis is applicable to glass fibre-polyester resin composites.

(ii) If the failure strain of the fibre < failure strain of the matrix (Fig. 6.13), at low values of f_f, when fibre fracture occurs the extra load is insufficient to fracture the matrix, so

$$\sigma_c^* = \sigma_m^* f_f + \sigma_m^* (1 - f_f) \tag{6.7}$$

When f_f is large, however, when fibre fracture occurs, the large extra load cannot be carried by the matrix and so the matrix fractures (at σ_m) when the fibres fail, giving:

$$\sigma_c^* = \sigma_f^* f_f + \sigma_m (1 - f_f) \tag{6.8}$$

The variation of σ_c^* with f_f is illustrated in Fig. 6.14, where f_{min} defines the volume fraction of fibres below which the fibre failure occurs below the matrix ultimate strength, and f_{crit}, above which the strengthening effect of the fibres is felt. This

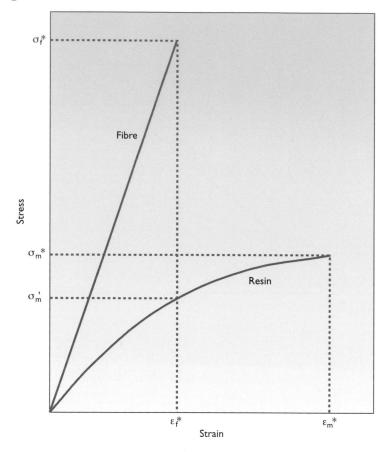

Fig. 6.13 Stress–strain curves of fibre and matrix with failure strain of matrix > that of fibre.

analysis can be applied to carbon fibre-epoxy resin composites.

The above approach has assumed that the fibre strength has a unique value, whereas for brittle fibres like glass, boron and carbon this is not the case, and their strengths are statistically distributed. A more sophisticated approach may be made to the problem, employing the Weibull statistics described in Chapter 2.

Aligned, Short Fibre Composites

If we assume the composite consists of an array of short fibres of uniform length (l) and diameter ($d = 2r$), all aligned in the loading direction and distributed uniformly throughout the material, then the matrix has the function of transferring the applied load to the fibres. The situation is rather like as in the model shown in Fig. 6.15, in which a thin fibre is embedded to a depth x in a medium which adheres to it. If the fibre is pulled, the adhesion between the fibre and the

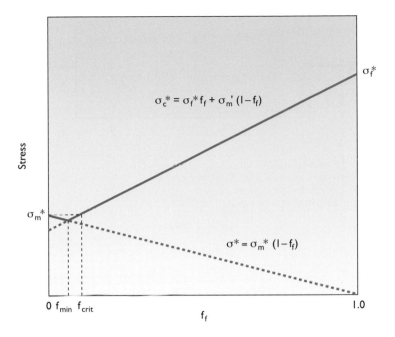

$$\sigma_c^* = \sigma_f^* f_f + \sigma_m' (1-f_f)$$

$$\sigma^* = \sigma_m^* (1-f_f)$$

Fig. 6.14 Variation of fracture stress of composite with volume fracture of fibres when $\epsilon_m^* > \epsilon_f^*$.

matrix produces a shear stress τ parallel to the surface of the fibre. The total force on the fibre is $2\pi r x \tau$ due to this shear stress. Suppose the maximum value of shear stress which this interface can withstand is $\tau_{\max}$, and the breaking stress of the fibre is σ_f^*: if we require that as we increase the pull on the fibre, the fibre shall break before it will pull out of the matrix, we must have

$$\pi r^2 \sigma_f^* < 2\pi r x \tau_{\max} \qquad (6.9)$$

i.e.

$$\frac{\sigma_f^*}{4\tau_{\max}} < \frac{x}{d}$$

The ratio x/d is called the aspect ratio of the embedded part of the fibre, and in order to fulfil this condition it is clear that the fibre has to be sufficiently long and thin.

Let us now consider a 'unit cell' (Fig. 6.16) of matrix containing a representative short fibre of length l. If a tensile load P is applied to each end of the unit cell, the load is transferred into the fibre from both ends by shear forces at the fibre/matrix interface. Consider an element of fibre at a distance x from one end, Fig. 6.17, balancing the loads on the element we have:

$$P - dP + 2\pi r \, dx.\tau = P$$

i.e.

$$dP/dx = 2\pi r.\tau$$

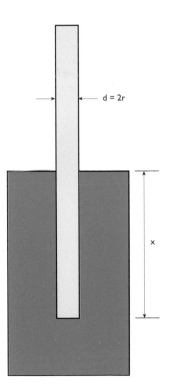

Fig. 6.15 Model of fibre embedded in an adherent medium.

$d = 2r$

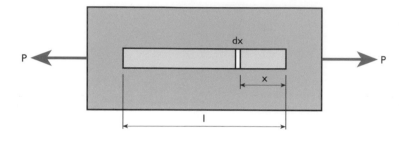

Fig. 6.16 Element of matrix containing a fibre.

Therefore $\qquad\qquad$ $P = 2\pi r\,\tau x$

If the applied stress upon the element is σ,

$$\pi r^2 \sigma = 2\pi r\tau x,$$

i.e. $\qquad\qquad$ $\sigma = 2\tau/r.x$

As shown in Fig. 6.18a, the stress in the fibre increases linearly from each end, reaching a maximum at the centre of the fibre, where $x = l/2$.

With increasing length of fibre, the maximum stress carried by the fibre increases until it reaches $\sigma_f{}^*$ when fibre failure will occur. The distance from one fibre end to the point of maximum stress is known as the 'transfer length' $(l/2)$, and for the whole fibre the *critical transfer length* (l_c) is required to achieve a stress of $\sigma_f{}^*$. Figure 6.18 shows the variation of tensile stress in a fibre as a function of fibre length: if the fibre is shorter than l_c, σ cannot reach the value for fibre fracture however much the composite is deformed. The critical aspect

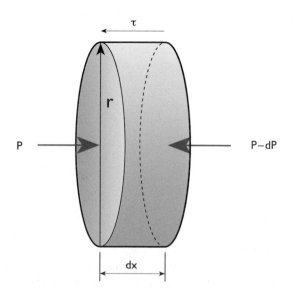

Fig. 6.17 Element of fibre in Fig. 6.16.

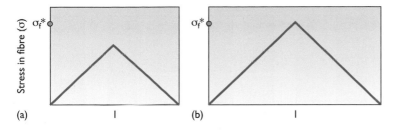

Fig. 6.18 Stress along fibre. a, $l < l_c$ b, $l = l_c$.

ratio for a fibre to be broken is thus:

$$\frac{l_c}{d} = \frac{\sigma_f{}^*}{2\tau_{\max}}$$

and only by using fibres much longer than l_c can the full strengthening potential of the reinforcement be achieved.

A fibre of length l_c will carry an average stress $(\bar{\sigma}_f)$ of only $\sigma_f{}^*/2$, and a general expression for the average fibre stress for a fibre of length l, is thus

$$\bar{\sigma}_f = \sigma_f{}^*(1 - l_c/2l)$$

and the composite strength, for $l > l_c$, calculated from the mixture rule on this basis is thus:

$$\sigma_c{}^* = \sigma_f{}^* f_f(1 - l_c/2l) + \sigma_m(1 - f_f) \qquad (6.10)$$

Equations 6.8 and 6.10 compare the strengths of continuous and short fibre composites, and it emerges that 95% of the strength of a continuous fibre composite can be achieved in a short-fibre composite with fibres only ten times as long as the critical length. On the other hand, it must also be remembered that in a panel made of a chopped fibre composite, the fibres may be randomly oriented in the plane of the sheet. Only a fraction of the fibres will be aligned to permit a tensile force to be transferred to them, so their contribution to the strength is correspondingly reduced.

The Fracture Behaviour of Composites

The Work of Fracture in Axial Tensile Parallel to the Fibres

If a rod of composite material containing a transverse notch or crack in the matrix (Fig. 6.19) is subjected to an axial force, the volume of material at the minimum cross-section will experience the greatest stress and so it will start to extend. This region will also suffer a lateral Poisson contraction, however, giving rise to transverse tensile stresses parallel to the plane of the notch. If a crack starts to run in a direction perpendicular to the fibres, the transverse tensile stress will cause the fibre/matrix interface to pull apart, or 'debond', as

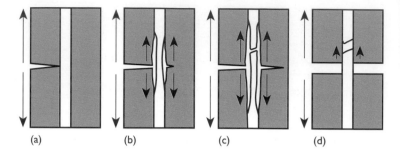

Fig. 6.19 Propagating transverse crack in an element of fibre composite.

(a)   (b)   (c)   (d)

shown in Fig. 6.19b. By this process the crack is deflected along the weak interface in the same way as when a notched stick of bamboo wood is bent. It does not snap into two pieces: it may splinter, but no cracks run into the material since its structure consists of strong bundles of cellulose fibres separated by lignin-based material at relatively weak-yielding interfaces.

Returning to Fig. 6.19, if the specimen is progressively further extended, the fibres across the crack face will bear the load and will eventually fracture. The component will eventually separate into two pieces by a process of fibre pull-out (Fig. 6.19d).

The total work of fracture of a fibre composite is thus composed of several terms, namely the work of debonding the fibre/matrix interface, the work of deformation of the matrix is the crack opens, and the work of fibre pull-out. In fibre composites with a brittle matrix, the latter term often dominates the toughness, and its magnitude may be estimated as follows for a composite containing a volume fraction f_f of fibres of length l $(>l_c)$.

When fracture occurs by the breaking of fibres, then all those fibres which end within a distance $l_c/2$ of the cross-section which crosses the break will pull out of the matrix.

The fraction pulling out is thus $= l_c/l$

Assuming the fibre/matrix interfacial shear strength (τ) is maintained constant:

The work to pull out one fibre of length $x =$

$$= \pi r^2 \int_0^x \sigma \, dx$$

$$= \pi r^2 \int_0^x (2\tau x/r) \, dx$$

$$= \pi r \tau x^2$$

For a volume fraction of fibres f_f, the number of fibres per unit area $= f_f/\pi r^2$.

In order to pull out, the fibres can only have a length *up to* $l_c/2$ above the plane of fracture: if this distance from the fibre end is x, then the work of pullout is exactly $\pi \, r \, \tau \, x^2$. (N.B. if the distance is $(x + dx)$, the work is still the same, since dx is vanishingly small).

Considering *all the possible positions* along the fibre where the plane of fracture might be:

The probability of it intersecting *exactly* at x is zero, but the probability of it intersecting *over the length dx* (i.e. of pulling out a length between x and $x + dx$ must be:

$$dx/\text{total length} = dx/\tfrac{1}{2}l_c$$

So the total work of pullout, W

$$= f_f/\pi r^2 . l_c/l \int_0^{l_0/2} \pi r \tau x^2 \; dx / (l_c/2)$$

$$= \frac{f_f \tau l_c^3}{12 \; r \; l}$$

But $l_c/r = \sigma_f/\tau_y$, so

$$\text{Work} = \frac{f_f}{12} \cdot \frac{l_c}{l} . \sigma_f l_c \qquad (6.11)$$

For maximum work of fracture, therefore, l_c should be maximised, and fibre length should be equal to l_c. If the fibre length is less than l_c, the work of fracture is obtained by setting $l_c = l$ in the integral:

$$\text{Work} = \frac{f_f}{\pi \; r^2} . \int_0^{\frac{l_c}{2}} \frac{(2\pi \; r \; \tau \; x^2)}{l} . dx$$

$$= f_f \; \pi . \tau l^2 / 12r. \qquad (6.12)$$

Thus for fibres of a given constant value of critical length, as the fibre length in the composite is increased, the work of pullout initially increases as l^2 (equation 6.12). When the fibre length is greater than the critical length, the work of pullout decreases as l^{-1}, (equation 6.11), as more fibres break rather than pull out. These relationships are illustrated in Fig. 6.20.

Fatigue Behaviour of Fibre Composites

The introduction of short fibres into a polymer can enhance its resistance to both the nucleation and the growth of fatigue cracks.

Thus Fig. 6.21 presents a series of $S–N$ curves for injection moulded PSF with increasing volume fractions of short glass

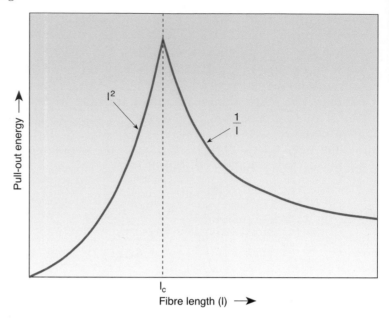

Fig. 6.20 Pull-out energy as a function of fibre length.

fibres, and with a volume fraction of 0.4 of carbon fibres, together with the S–$\mathcal{N}$ curve for the matrix material itself. It is clear that the introduction of the fibres progressively raises the endurance limit, and leads to a increased fatigue life at all stress amplitudes.

In Fig. 6.21, the higher stiffness of the carbon fibres promotes a higher fatigue life still, by reducing the cyclic strains in the matrix for a given stress amplitude. It has also been observed that the high thermal conductivity of carbon reduces the temperature rise due to hysteretic heating, although all fibre strengthening of polymers is reduced at high test frequencies, when thermal softening controls the fatigue life.

With regard to crack propagation, Fig. 6.22 shows a series of da/dN versus ΔK plots, for an unreinforced polymer and also for this matrix reinforced with 20% and 30% glass fibres. The fibres are seen to confer considerable resistance to the propagation of fatigue cracks to an extent which increases with their volume fraction. It is also apparent that the effect of the fibres is greater for cracks growing normal to the mould fill direction than for those growing along the mould fill direction.

The fatigue behaviour of a metal matrix composite is also usually superior to that of the unreinforced matrix, particularly in the case of unidirectional fibre composite system. Three types of S–$\mathcal{N}$ curves have been reported, as illustrated in Fig. 6.23, whose shape depends upon the mode of crack

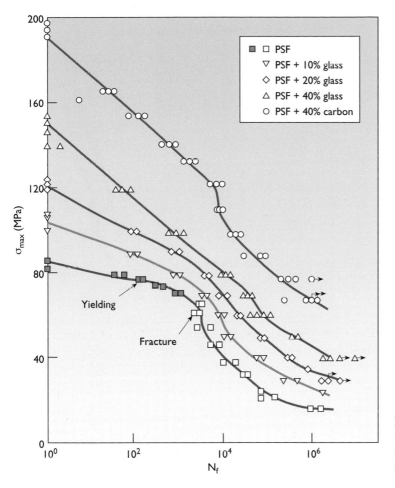

Fig. 6.21 S–N curves for PSF reinforced with different amounts of short fibres of glass or carbon.

propagation. If the matrix crack grow in a direction perpendicular to the fibre axis, the sigmoidally-shaped S–N curve marked A is observed. If interfacial debonding takes place by the crack progressing in a direction parallel to the fibres, if the fibres are perfect then the endurance limit will correspond to the static strength and the S–N curve will be horizontal (marked B). The majority of fibres will not be perfect, however, but will possess local weak points, so that a mixed type of fracture occurs and the S–N curve will be a mixture of types A and B (Fig. 6.23).

Environmental Effects in Fibre-Reinforced Composites

We have already discussed environmental effects upon metals and upon polymers. Most practical reinforcing fibres (with the possible exception of the aromatic polyamides such as Kevlar) are unlikely to be affected by ultraviolet radiation, and since the bulk of the fibres in the composite are in any

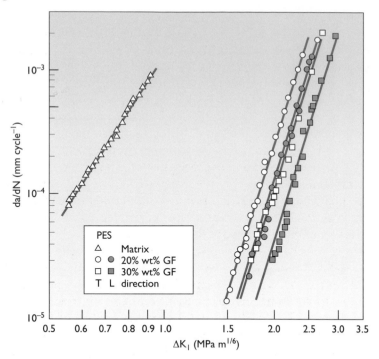

Fig. 6.22 Long $(\mathrm{d}a/\mathrm{d}N)$ log (ΔK) curves of PES composites tested at a frequency of 5Hz.

case protected from the radiation by the resin, any degradation from this cause is likely to be dominated by effects upon the matrix. Carbon fibre may have a similar effect as carbon black in absorbing ultraviolet radiation, and will alleviate the effect of radiation damage to the polymer by converting it to thermal energy.

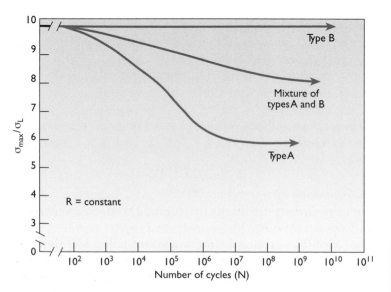

Fig. 6.23 Illustration of three types of S–N curves for MMCs.

In MMCs, aqueous environments may lead to severe corrosion problems if the metal/fibre combination forms a galvanic couple. Carbon fibre – aluminium composites have been identified as been prone to this type of attack. The principal attack occurs where fibres are exposed, at cut or machined surfaces for example. Similar problems have been encountered when carbon fibre reinforced plastics have been used in conjunction with metallic components (such as fixing screws) in the presence of an aqueous environment.

Moisture can cause degradation of properties in fibre-reinforced polymers if it is transported to or diffuses along the fibre/matrix interface. It may cause breakdown of bonding between fibre and matrix, thus impairing the efficiency of stress transfer in the composite. If the fibre/matrix bonding is purely mechanical in nature, then the properties of the composite would be expected to recover when it is dried out, but if the bonding is chemical, then unless these bonds are restored on drying, permanent damage will result.

Glass and Kevlar-49 fibres are themselves affected by quite low levels of exposure to moisture, being susceptible to stress-corrosion cracking, and the effect is even more pronounced in acid or alkaline environments. Extensive stress-rupture tests upon a range of GRP with different resins, in aqueous environments at various temperatures have confirmed that the strength of moist GRP is controlled by moisture sensitivity of the fibres, although different resins result in different lives for a given stress level.

The Mechanical Properties of Cellular Solids

The properties of a foam are described in terms of their *relative density* , ρ/ρ_s, where ρ is the density of the foam and ρ_s that of the solid of which it is made. This ratio can vary over two orders of magnitude with foams of differing density.

Elastic Properties

At small strains, foams behave in a linear elastic manner, and their modulus is given by

$$E = E_s \left(\frac{\rho}{\rho_s} \right)^n$$

where E_s is the modulus of the solid. It is found experimentally that in tension $n \approx 1.5$ and $n \approx 2$ in compression. E can vary over four orders of magnitude with foams of different relative density. This equation may also be employed to describe the properties of wood, in which case ρ is the density of the cell wall material. For moduli parallel to the grain, $n = 1$, which corresponds to a rule of mixtures, and for the perpendicular direction $n = 2$ is a better approximation.

At larger strains foams deform by elastic buckling of the walls, and from standard beam theory, the elastic collapse stress, σ_{el}^*, is given by

$$\sigma_{el}^* = 0.05 \; E_s \left(\frac{\rho}{\rho_s} \right)^2$$

If the matrix itself is capable of plastic deformation, then in cellular form it may exhibit plastic collapse, which is non-reversible but which will absorb considerable energy if the material is used for padding purposes, as in the case of polyurethane automobile crash padding. If the yield strength of the solid matrix is σ_y, then the collapse stress, σ_{pl}^*, is given by

$$\sigma_{pl}^* = 0.3 \; \sigma_y \left(\frac{\rho}{\rho_s} \right)^{3/2}$$

READING LIST

Bryan Harris: *Engineering Composite Materials,* 2nd edition, London: The Institute of Materials 1999.

Derek Hull: *An introduction to composite materials,* 2nd edition, Cambridge University Press 1996.

Minoru Taya & Richard J. Arsenault: *Metal matrix composites – thermomechanical behaviour*, Oxford, Pergamon Press 1989.

General Reference

Engineered Materials Handbook, Volume 1: Composites, Ohio: ASM International 1987

PROBLEMS

CHAPTER I

1. Distinguish between the following types of bonding in solids:
   van der Waals bonds
   covalent bonds
   ionic bonds
   metallic bonds.
Name four materials in which each of these types of bonding occur.

2. Show that differentiation of equation (1.4) leads to equation (1.5). Obtain an expression for the value of ΔG^* (Fig. 1.5), which is the activation free energy for the formation of a nucleus of critical size (r_c). By substituting equation (1.3) into your expression, derive an equation which describes the dependence of the value of ΔG^* upon the degree of supercooling.

   Can your derived equations enable you to predict qualitatively the observed relative grain sizes of metal objects formed by casting the molten metal (a) into a cold metal mould of large wall thickness, and (b) into a preheated sand mould?

   Estimate the critical nucleus size and the free energy of formation of the critical nucleus of solid tin at 473 K given the following data
   Melting point of tin $= 505$ K
   Solid liquid interfacial energy $= 0.1$ J m^{-2}
   Latent heat of solidification $= 4 \times 10^5$ KJ m^{-3}.

[7.9 nm]

3. Discuss the appropriate metallographic techniques for the following studies:
  (a) determination of the grain size of a sand-cast bronze marine propeller blade.
  (b) determination of the grain size of a mild sheet plate.
  (c) examination of a fatigue fracture surface.
  (d) the identification of a precipitated phase in a tempered steel.

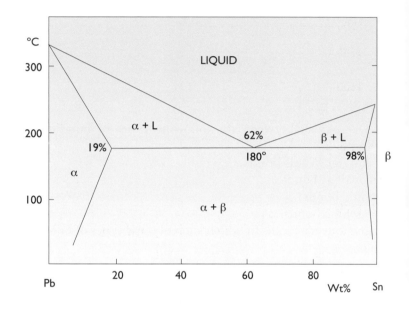

4. During the solidification of an alloy, it is assumed that there is complete mixing in the liquid and no mixing in the solid (i.e. 'cored' crystals are formed). The instantaneous concentration of the solid in contact with the liquid phase, C_x, may be expressed in terms of the fraction frozen x:

$$C_x = C_0 k(1 - x)^{k-1}$$

where C_0 is the initial liquid concentration and k is the solid/liquid distribution coefficient (i.e. the ratio of solute in the solid phase to that in the liquid phase at a given temperature - it may be assumed to be constant).

Assuming that this equation applies to a casting, estimate the proportion of non-equilibrium second phase which would occur in an alloy of Pb-10 wt%Sn, the equilibrium diagram of which is shown above.

[Answer: A fraction of 0.072].

5. With reference to the Ti-Ni equilibrium diagram reproduced below, discuss the reactions and the probable microstructures formed when the following binary alloys of titanium are slowly cooled from 1600°C to 800°C:

(a) 10 weight % nickel
(b) 50 weight % nickel
(c) 70 weight % nickel.

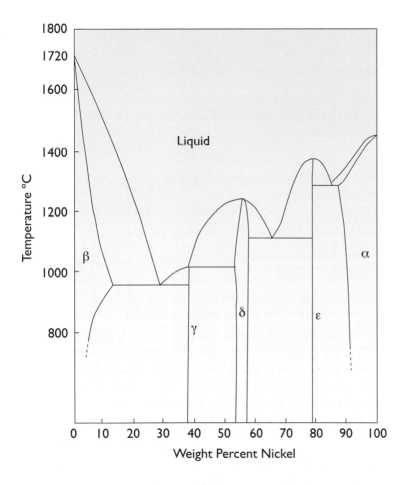

6. The molecular weight distribution was determined for a polyethylene sample. The following number fractions (N_i) corresponding to particular molecular weight ranges (M_i) were measured:

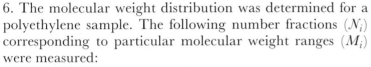

N_i 0.26   0.31   0.21   0.13   0.07   0.015 0.001

M_i 10^3 3×10^3 10^4 3×10^4 10^5 3×10^5 10^6

Calculate the *number average* molecular weight and the *weight average* molecular weight for this polymer.

7. Sketch portions of a linear polypropylene molecule that are (a) isotactic, (b) syndiotactic and (c) atactic.

8. Compare the crystalline state in metals and polymers.

9. Compare the non-crystalline state as it applies to polymers and inorganic glasses.

CHAPTER 2

1. The portion of the true stress–strain curve for a certain material may be described empirically by the Hollomon relationship:

$$\sigma = K\epsilon^n$$

where n is known as the strain hardening coefficient.

In a tensile test on this material, find the value of the true plastic strain at necking instability.

$[n]$.

2. Find the maximum surface stress in an elastically loaded beam of width 10 mm and thickness 5 mm under a load of 2 Kg in (a) three-point bending with 100 mm between the outer knife edges (L) and (b) four-point bending with the 25 mm between the inner and outer knife-edges (d).

[(a) 120 MPa; (b) 10 MPa]

3. Equation 2.9 gives the Weibull distribution for the probability of survival in the fracture testing of brittle solids, where σ_0 is mean stress at failure for a given material. We may define $P_s = n/(N+1)$, where N is the total number of samples.

A set of ten square ceramic bars was tested in tension with the following results for the fracture strengths measured in MPa:

Bar number(n)	1	2	3	4	5	6	7	8	9	10
Fracture strength	170	160	159	154	143	135	128	118	105	100

(a) What is the value of σ_0?

When the volume of the specimens, $V = V_0$, what is the probability of survival (P_s) when:
  (b) $\sigma = \sigma_0$
  (c) $\sigma = 2\sigma_0$
  (d) $\sigma = \frac{1}{2}\sigma_0$
  (e) What is the Weibull modulus (m) for the given ceramic?

[(a) 137.2 MPa; (b) 0.368; (c) 1.3×10^{-14} (d) 0.969; (e) 5.419]

4. A plate of brittle material contains a crack of length 1 mm. It fractures under a tensile stress, applied normal to the crack, of 600 MPa. Estimate the apparent surface energy of the material, and comment on the magnitude of your result. (Young's Modulus of the material $= 200$ GPa).

[1414 J m^{-2}]

5. Calculate the minimum size of defect in a thin plate of marageing steel that would give rise to brittle fracture when a stress equal to $\frac{2}{3}$ of the proof stress is applied. (For the marageing steel, proof stress $= 1200$ MPa, Young's modulus $= 200$ GPa and $G_c = 24$ kJ m^{-2}) ($Y = $ unity in equation 2.13a).

[6.2 mm].

6. The creep strain ϵ, measured as a function of time t for a nickel specimen at 1000 K may be fitted to a curve of the form illustrated in Fig. 2.11 (A). Neglecting the initial instantaneous elastic strain, the data may be fitted to the following expression:

$$\epsilon = at^{1/3} + bt + ct^2$$

Explain, giving experimental details, how the coefficients a, b and c can be measured.

[If the strain, the strain rate and the time are measured at the point at which the creep strain rate is a minimum, then:

$$a = \frac{9(\varepsilon - \dot{\varepsilon}t)}{5t^{1/3}}; b = \frac{2\dot{\varepsilon}t - \varepsilon}{t}; c = \frac{\varepsilon - \dot{\varepsilon}t}{5t^2}]$$

7. A specimen in a tensile creep test at 800°C ruptures after

1000 hours. If the activation energy for the dominant creep mechanism under these test conditions is 200 kJ mol^{-1}, estimate the time to rupture of a similar specimen under the same stress at 600°C, discussing any assumptions you make.

[1.6×10^5 hours].

8. A nickel alloy bolt is used to hold together furnace components. The bolt is initially tightened to a stress σ_1, and must be re-tightened if the stress falls below a critical level σ_c. The alloy creeps according to equation (2.15): derive an expression for the time required between re-tightenings.

Calculate the maximum time between re-tightenings at 500°C if the bolt must not exceed $\frac{1}{3}$ of its yield stress of 600 MPa, and if σ_c is taken to be 100 MPa. In an attempt to extend the time between re-tightenings, the safety factor is ignored and the bolt is tightened to 60% of its yield stress. What re-tightening time will this give?

(For this alloy, $n = 4.6$, σ_0 is unity, $Q = 270$ kJ mole^{-1}, $\dot{\epsilon}_0 = 8 \times 10^{-36}$ s^{-1} and Young's modulus is 215 GPa).

[379 days and 46 days]

9. Using equation (2.16), calculate the number of cycles for fatigue failure (N_f) when the following materials are subject to a cyclic stress amplitude equal to 0.8 of their yield stress: normalised 0.15% carbon steel, peak aged 7075 AlZnMg alloy and a quenched and tempered NiCrMo 4340 steel.

[4.65×10^5 cycles; 1.064×10^5 cycles; 833 cycles]

10. The rate of fatigue crack growth in a certain alloy follows the Paris power law (equation (2.24)). A wide plate of this material failed by growth of a fatigue crack from one edge under a constant stress amplitude. Fatigue striations were 2 μm apart on the fracture crack surface at a distance of 4 mm from the edge and 5.5 μm apart at a distance of 9 mm from the edge. Determine the value of m in equation (2.24), given that Y equals unity in equation (2.23).

[2.49]

11. Calculate the energy dissipated per cycle of stress when

the following materials are subjected to elastic vibrations at a stress amplitude of 31 MPa:

(a) mild steel of Young's modulus 200 GPa and specific damping capacity 2.28.

(b) cast iron of Young's modulus 170 GPa and specific damping capacity 28.

Which material would be more appropriate for use for the manufacture of a lathe bed, and why?

[(a) 5.48 kJ; (b) 79.1 kJ]

12. At a temperature close to its glass transition, a certain cross-linked amorphous polymer is found to deform under uniaxial stress, σ, according to a Voigt model (of Young's modulus E_1 and viscosity η) in series with a spring of Young's modulus E_2. Determine the stress σ as a function of time (t) when a constant strain ϵ is instantaneously applied to the specimen.

$[\sigma = \epsilon[E_R + (E_2 - E_R)\exp(-t/\tau)],$
where $E_R = 1/(1/E_1 + 1/E_2)$, and $\tau = \eta/(E_1 + E_2).]$

CHAPTER 3

1. The tensile yield strength of annealed iron (grain size 16 grains mm^{-2}) is 100 MPa, and 250 MPa for a specimen with small grain size (4,096 grains mm^{-2}). Determine the yield strength of iron with a grain size of 250 grains mm^{-2}.

[149.4 MPa]

2. Describe how a recrystallization treatment may be used to prepare a single-phase alloy:

(a) with a very fine grain size
(b) with a coarse grain size.

3. Aluminium containing a given volume fraction of (unshearable) oxide particles of average diameter 200 nm has a shear yield strength of 30 MPa. What would be the shear yield strength if the microstructure consisted of a dispersion of the same volume fraction of 40 nm diameter oxide particles?

[150 MPa]

4. 316 stainless steel with a grain size of 50 μm and at a temperature of 1090 K deforms at a shear strain rate $\dot{\gamma}$ which varies with shear stress τ in the following way:

τ (Pa)	6.1×10^4	6.1×10^5	6.1×10^6	3.7×10^7
$\dot{\gamma}$ (s^{-1})	10^{-10}	10^{-9}	10^{-8}	10^{-7}

5.1×10^7	6.6×10^7	9.2×10^7	1.2×10^8	1.6×10^8
10^{-6}	10^{-5}	10^{-4}	10^{-3}	10^{-2}

Explain this behaviour in as much detail as you can, identifying the creep mechanisms operating.

5. A component is exposed to fluctuating stresses in service: in order to minimise the danger of it failing by fatigue discuss the steps that should be taken (a) in the *design* of the component and (b) in the *selection* of the material itself and any *special treatment* it may be given.

6. Compare the response to heat-treatment of a Ti-2.5Cu alloy with that of IMI318.

7. Sketch on the same axes three hypothetical true stress–strain curves of the shapes you would expect for a medium carbon steel which has been quenched and then tempered at (a) 350°C, (b) 450°C and (c) 550°C.
   Construct a Considère tangent to each curve (Fig. 2.5). In the light of your construction discuss the relationship between the tempering temperature and the per cent elongation to tensile fracture for this material (Fig. 3.24).
   On your diagram construct another hypothetical true stress–strain curve for a material which would be expected to exhibit both high strength *and* high ductility.

8. Figure 3.16 indicates that α-brasses (containing up to 30% of zinc) exhibit a progressive increase in both tensile strength and in the elongation to fracture as the zinc content increases. This contrasts with the behaviour of quenched and tempered steels (Fig. 3.24) where the tensile elongation *decreases* as the strength rises. Can you give an explanation for this difference in behaviour between brass and steel?

9. A welded joint is produced between two cold-rolled sheets of type 304 austenitic stainless steel. Sketch a diagram for this situation analogous to that on the left-hand side of Fig. 3.32.

Discuss any metallurgical changes that may take place in the heat-affected zone, and any problems that might be encountered with this joint when the component is in service in a corrosive environment. How might such problems be averted?

10. Give an account of the factors which give rise to an iron casting being 'white' or 'grey' in character. What methods are available for producing ductile iron castings?

11. Give an account of the factors which determine whether or not an oxide film which has been formed on the surface of a metal or alloy protects it from further oxidative attack. To what extent may alloys be designed to exhibit oxidation resistance?

12. Two mild steel plates are to be rivetted together for subsequent use in a structure to be in contact with seawater. Rivets made of four materials are available namely (a) stainless steel, (b) copper, (c) carbon steel and (d) aluminium. Each of these types of rivet is tried, and each riveted structure is painted before being put into service.

Discuss the nature of any corrosion problems that may be encountered in each of the four riveted assemblies.

13. Extracting the data from Fig. 3.36, estimate the dimensional wear coefficient (k) for a leaded α/β brass sliding against a hard stellite ring (a) in the region of mild wear and (b) in the region of severe wear.

Discuss the mechanisms of wear in these two regimes.

$[(a)1.25 \times 10^{-5}; (b)3.17 \times 10^{-4}]$.

CHAPTER 4

1. Silicon carbide and silicon nitride have respective thermal expansion coefficients $(0-1000°C)$ of 4.5×10^{-6} K^{-1} and 3.0×10^{-6} K^{-1}, and thermal conductivities (at $100°C$) of 125 Wm^{-1} K^{-1} and 20 Wm^{-1} K^{-1}.

Discuss their relative ability to withstand thermal shock.

2. A ceramic of Young's modulus 200 GPa and effective surface energy for fracture of 40 Jm^{-2} contains a large flaw 100 μm in size. If its thermal expansion coefficient is 3×10^{-6} K^{-1} over a substantial range of temperatures, estimate the maximum temperature change which the ceramic, in the

form of a bar, can withstand if its ends are held fixed while the bar is cooled.

[470°C]

3. Pores of diameter 5 μm are sealed in a glass under nitrogen at a pressure of 8×10^4 Pa and the glass is then annealed in vacuum. Assuming that nitrogen is completely insoluble in glass, determine the final (equilibrium) pore size if the surface energy of the glass is 0.3 Jm^{-2}.

[2.8 μm].

4. The measured strengths of a certain ceramic are believed to follow the Weibull distribution. From an extensive series of tests on identical specimens the probability of surviving a stress of 700 MPa is estimated to be 0.9 and that of surviving 900 MPa, 0.1. Determine the Weibull modulus for the ceramic and the design stress for a 99% survival probability for a ceramic component having a volume 10 times as great as the volume of the specimens tested in establishing the above survival probabilities.

[$m = 12.275$; 479 MPa]

5. A concrete casting is of Young's modulus 50 GPa. When normally prepared it contains voids of diameter up to 1 mm; if the void size were reduced to a maximum of 15 μm in size, what change in tensile fracture stress might be expected if the fracture surface energy were 1 Jm^{-2}?

[About +40 MPa].

6. Calculate the value of Young's modulus of concrete containing the following volume fractions of aggregate (a) 0.45, (b) 0.6 and (c) 0.75.

[48.4 GPa; 58.4 GPa; 73.6 GPa].

7. Outline the basic chemistry of the microstructural changes occurring during the setting and hardening of Ordinary Portland cement.

CHAPTER 5

1. What properties do you expect in covalently bonded solids?

How do you account for the differences in mechanical behaviour between polythene and diamond in both of which the carbon atoms are covalently bound?

2. Compare and contrast the properties and uses of organic, inorganic and metallic glasses.

3. If a strain of 0.4 is applied to an elastomer, an immediate stress (σ_0) of 8 MPa is produced which instantly relaxes as elastic strain is replaced by viscous strain, so that after a time (t) of 42 days the stress (σ) is only 4 MPa. This behaviour may be described by the expression

$$\sigma = \sigma_0 e^{-t/\tau}$$

where τ is the *relaxation time* of the material.

What is the relaxation time of the above elastomer, and what would the stress be after 90 days?

[60.1 days; 1.7 MPa]

4. Compare and contrast the deformation in a tensile test of a ductile polymer such as high density polyethylene with that of a ductile metal. Discuss the effects of extreme tensile deformation of the properties of an initially isotropic polymer, and discuss whether such effects may be of practical value.

5. The *melt viscosity* η of high molecular weight atactic polystyrene is given by the following two relationships:

$$\eta = K_H Z_W^{3.4}$$

and

$$\log_{10}(\eta_T / \eta_{T_s}) = \frac{17.44(T - T_g)}{51.6 + (T - T_g)}$$

where K_H is a constant, Z_w is the weight-averaged number of backbone atoms per molecule, and T is the temperature. T_g is the glass transition temperature of $100°C$.

An injection moulding unit normally produces polystyrene drinking cups at $160°C$ (where the melt viscosity is 1500 poise), using polymer for which $Z_W = 800$. Due to a problem with the supplier it becomes necessary to use temporarily a polymer for which $Z_W = 950$.

At what temperature should this new batch of polymer be moulded?

6. Give an account of the phenomenon of crazing in poly-styrene.

7. Give an account of elastomer toughening of polystyrene.

8. Compare the phenomenon of fatigue failure in polymeric materials with that in metallic materials.

9. Compare the degradation of polymeric materials and of metallic materials which arises from their interaction with their environment.

CHAPTER 6

1. Calculate the value of Young's modulus for a cemented carbide cutting tool consisting of a uniform array of cube-shaped particles of tungsten carbide (Young's modulus 703.4 GPa) in a matrix of cobalt (Young's modulus 206.9 GPa), if 60 volume % of the particles are present.

[see Figure 6.6]

2. A model fibre-composite system is prepared consisting of a uniform array of continuous tungsten wires in a copper matrix. If the wires occupy 50% of the cross-section, what will be the value of Young's modulus for the composite specimen? (The data required are on page 213).

[270.4 GPa]

3. A directionally solidified eutectic alloy has a microstruc-ture consisting of alternate continuous lamellae of phase α (of Young's modulus 70 GPa) and phase β (of Young's modulus 150 GPa) - the volume fraction of β-phase being 0.4.
   The value of Young's modulus of the alloy is measured with the tensile axis being first parallel to the lamellae and then with the tensile axis perpendicular to the lamellae. Calculate the degree of elastic anisotropy of the material.

[0.146].

4. A unidirectional fibre composite consists of a volume fraction of 0.6 of continuous carbon fibres in a matrix of epoxy. Find the maximum tensile strength of the composite if the matrix yields in tension at a stress of 40 MPa and the fracture strength of the fibres is 2700 MPa.

[1636 MPa]

5. A composite material consists of a mixture of short glass fibres (fracture strength 2000 MPa) of diameter 15 μm, in a polyester matrix whose shear strength is 30 MPa. What is the critical fibre length (l_c) for this system? Estimate the maximum work of fracture of the composite. How would you expect the work of fracture to change if the fibres were substantially longer than l_c?

[0.5 mm; 25 kJ m^{-2}]

6. Calculate the value of Young's modulus (a) parallel to the grain, and (b) perpendicular to the grain for a softwood of density 550 kg m^{-3}. The value of Young's modulus of cellulose is 40 GPa, and its bulk density can be taken as 1500 kg m^{-3}.

[(a) 14.6 GPa; (b) 5.4 GPa]

7. Explain why the interfacial shear stress is important in controlling the fracture behaviour of fibre reinforced composites.

Some Useful Constants

	Symbol	SI units
Absolute zero temperature		$-273.2°C$
Acceleration due to gravity	g	9.81 m s^{-2}
Avogadro constant	N_A	6.023×10^{23}
Base of natural logarithms	e	2.718
Boltzmann constant	k	$1.38 \times 10^{-23} \text{ J K}^{-1}$
		$8.62 \times 10^{-5} \text{ eV K}^{-1}$
Electron charge	e	$1.60 \times 10^{-19} \text{ C}$
Electron mass	m_e	$9.11 \times 10^{-31} \text{ kg}$
Faraday constant	F	$9.648 \times 10^4 \text{ C mol}^{-1}$
Gas constant	R	$8.31 \text{ J K}^{-1} \text{ mol}^{-1}$
Permeability of free space	μ_0	$4\pi \times 10^{-7} \text{ H m}^{-1}$
Permittivity of free space	ϵ_0	$8.85 \times 10^{-12} \text{ F m}^{-1}$
Planck constant	h	$6.63 \times 10^{-34} \text{ J s}$
Proton mass	m_p	$1.67 \times 10^{-27} \text{ kg}$
Velocity of light *in vacuo*	c	$3 \times 10^8 \text{ m s}^{-1}$
Volume of perfect gas at STP		$22.41 \times 10^{-3} \text{ m}^3 \text{ mol}^{-1}$

Unit Conversion Factors

LENGTH

1 Å	=	10^{-10} m
1 in	=	25.4 mm
1 ft	=	0.3048 m
1 yd	=	0.9144 m

VOLUME

1 Imperial gall.	=	4.546×10^{-3} m^3
1 US gall.	=	3.785×10^{-3} m^3

ANGLE

1 radian	=	$57.3°$

MASS

1 lb	=	0.4536 kg
1 short ton	=	907 kg
1 long ton	=	1016 kg
1 tonne	=	1000 kg

DENSITY

1 lb ft^{-3}	=	16.03 kg m^{-3}

FORCE

1 dyne	=	10^{-5} N
1 lb force	=	4.448 N
1 kg force	=	9.807 N

STRESS

1 psi	=	6.9×10^{-3} MPa
1 long tsi	=	15.44 MPa
1 dyne cm^{-2}	=	10^{-7} MPa
1 bar	=	0.1 MPa

FRACTURE TOUGHNESS

1 psi $\sqrt{\text{in}}$ $=$ 1.099×10^{-3} MPa m$^{\frac{1}{2}}$

ENERGY

1 erg	$=$	10^{-7} J
1 eV	$=$	1.602×10^{-19} J
1 cal	$=$	4.187 J
1 Btu	$=$	1054 J
1 ft-lbf	$=$	1.356 J

POWER

1 hp	$=$	0.746 kW
1 erg s^{-1}	$=$	10^{-10} kW
1 ft lbf s^{-1}	$=$	1.36×10^{-3} kW

SURFACE ENERGY

1 erg cm^{-2} $=$ 1 mJ m^{-2}

VISCOSITY

1 poise $=$ 0.1 N.s m^{-2}

SPECIFIC HEAT

1 cal g^{-1} K^{-1} $=$ 4.188 kJ kg^{-1} K^{-1}

PREFIXES

T	tera $= 10^{12}$		m	milli $= 10^{-3}$
G	giga $= 10^{9}$		μ	micro $= 10^{-6}$
M	mega $= 10^{6}$		n	nano $= 10^{-9}$
k	kilo $= 10^{3}$		p	pico $= 10^{-12}$
c	centi $= 10^{-2}$		f	femto $= 10^{-15}$

Selected Data for Some Elements

Element	E GPa	μ GPa	Density at 20°C 10^3 kg m^{-3}	Crystal structure at 20°C	Lattice parameter* nm	m.p. °C
Al	70.6	26.2	2.7	fcc	0.404	660.4
Be	318	156	1.848	hcp	0.228; 0.357	127.8
Cr	279	115.3	7.1	bcc	0.289	1875
Cu	129.8	48.3	8.96	fcc	0.361	1084
Fe	211.4	81.6	7.87	bcc	0.286	1538
Pb	16.1	5.59	11.68	fcc	0.495	327
Mg	44.7	17.3	1.74	hcp	0.321; 0.521	649
Mo	324.8	125.6	10.2	bcc	0.314	2617
Ni	199.5	76	8.9	fcc	0.352	1453
Pt	170	60.9	21.45	fcc	0.392	1772
Si	113	39.7	2.34	diamond cubic	0.542	1410
Sn	49.9	18.4	7.3	tetragonal	0.583; 0.318	232
Ti	120.2	45.6	4.5	hcp	0.295; 0.468	1668
W	411	160.6	19.3	bcc	0.316	3410
Zn	104.5	41.9	7.14	hex	0.266; 0.494	420
Zr	98	35	6.49	hcp	0.323; 0.514	1852

*Corrected values c.f. the first edition.

Sources of Material Property Data

HANDBOOKS

Metals

1. *ASM Metals Handbook*, 10th Edition (1990), ASM International, Metals Park, Columbus, Ohio, USA.
2. *Smithells Metals Reference Book* 7th Edition (1992)(Ed. E.A. Brandes), Butterworth Heinemann, London UK.

Ceramics and Glasses

1. *ASM Engineered Materials Handbook*, Vol. 4: Ceramics and Glasses (1991), ASM International, Metals Park, Columbus, Ohio, USA.
2. *Handbook of Ceramics and Composites*, (1990) 3 Vols, (Ed. N.P. Cheremisinoff), Marcel Dekker Inc., New York, USA.

Polymers and Elastomers

1. *ASM Engineered Materials Handbook*, Vol. 2: Engineering plastics (1989), ASM International, Metals Park, Columbus, Ohio, USA.
2. *Handbook of elastomers*, (1988), A.K. Bhowmick and H.L. Stephens, Marcel Dekker, New York, USA.

Composites

1. *Engineers Guide to Composite Materials*, (1987), Eds J.W. Weeton, D.M. Peters and K.L. Thomas, ASM International, Metals Park, Columbus, Ohio, USA.
2. *ASM Engineered Materials Handbook*, Vol. 1: Composites (1987), ASM International, Metals Park, Columbus, Ohio, USA.

DATABASE SOFTWARE

There is a rapidly increasing number of computer-based materials information systems available, and only a small selection of them is given here. The Materials Information Service of the Institute of Materials publish an updated list of PC Based Materials Databases, from whom copies are available.

Materials property databases can be either on- or off-line, but only the latter are considered below, as they are usually easier to use by the engineer.

Metals

1. MATUS is a series of diskettes for the PC produced by the **Engineering Information Company** in conjunction with the DTI and a variety of trade associations and manufacturers.

2. SOCRATES, produced by **Cortest Laboratories**, is designed to help selection of corrosion resistant alloys.

3. **TWI** group have produced a database of fracture toughness properties, and a bibliographic CD-ROM database on Surface Finishing.

4. Data on individual metals and alloys are also available as follows:

The Copper Development Association:
   Copper and copper alloys.
Magnesium Elektron Ltd:
   Magnesium alloys.
Nickel Development Institute:
   Stainless steels.
Titanium Information Group:
   Titanium and its alloys.
MPR Publishing Services Ltd:
   Powder Metallurgy (PM) Selector.
Aluminium Federation:
   'Aluselect'.

Ceramics

1. The NIST 'Structural Ceramics Database', produced by the **National Institute of Standards and Technology** (USA), consists of information on the thermal, mechanical and corrosion properties of silicon carbides and silicon nitrides.

2. The NIST 'Phase Diagrams for Ceramics and Database' is also available.

Polymers and Elastomers

1. PLASCAMS, produced by **Rapra Technology Ltd**, is designed to take the user from an initial specification through to a suitable material and supplier (polymers only).

2. CHEMRES, also produced by **Rapra Technology Ltd**, is designed to select plastics according to their chemical resistance.

3. *data*PLAS, produced by **Polydata** for engineers in the US and Canada, contains information on 1000 high-performance thermoplastics.

4. CAPS and CAMPUS, also produced by **Polydata**, are respectively a system for the selection of thermoplastic grades, and a product information system operated by about 30 suppliers of raw materials.

5. MORPHS, produced by **Rubber Consultants**.

6. RUBBACAMS, is a Rubber Materials Selector produced by **Rapra Technology**.

7. INTERNATIONAL PLASTICS SELECTOR, produced by **HTI Ltd**.

Adhesives

1. PAL, produced by **Permabond Adhesives Ltd**.

Multi-Material Data Bases

1. M/VISION, produced by **PDS Engineering**, requires a workstation, and may be integrated with CAD and CAE systems. It contains data for aerospace alloys and composites.

2. The Cambridge Materials Selector, produced by **Cambridge University Engineering Department**, allows the rapid selection and evaluation of materials for design. It includes data for metals, polymers, ceramics, composites and natural materials.

3. The basic program of Mat.D.B., produced by **ASM International**, is the database 'engine' plus a small starter database. Additional in-depth databases may be purchased separately for a wide range of materials.

4. PERITUS, produced by **Matsel Systems Ltd**, is a database for metals, polymers and ceramics, aimed at materials and process selection.

5. METADEX, produced by **Dialog Europe** of Oxford.

6. LAMINATE ANALYSIS PROGRAMME, produced by the **Centre for Composite Materials**, Imperial College of Science and Technology, London.

The Periodic Table of the Elements

Legend:
- At.Wt
- Element
- At.No.

IA	IIA	IIIB	IVB	VB	VIB	VIIB	VIII	VIII	VIII	IB	IIB	IIIA	IVA	VA	VIA	VIIA	
1 H 1																	4 He 2
7 Li 3	9 Be 4											11 B 5	12 C 6	14 N 7	16 O 8	19 F 9	20 Ne 10
23 Na 11	24 Mg 12											27 Al 13	28 Si 14	31 P 15	32 S 16	35.5 Cl 17	40 Ar 18
39 K 19	40 Ca 20	45 Sc 21	48 Ti 22	51 V 23	52 Cr 24	55 Mn 25	56 Fe 26	59 Co 27	59 Ni 28	64 Cu 29	65 Zn 30	70 Ga 31	73 Ge 32	75 As 33	79 Se 34	80 Br 35	84 Kr 36
85.5 Rb 37	88 Sr 38	89 Y 39	91 Zr 40	93 Nb 41	96 Mo 42	98 Tc 43	101 Ru 44	103 Rh 45	106 Pd 46	108 Ag 47	112 Cd 48	115 In 49	119 Sn 50	122 Sb 51	128 Te 52	127 I 53	131 Xe 54
133 Cs 55	137 Ba 56	139 La 57	178.5 Hf 72	181 Ta 73	184 W 74	186 Re 75	190 Os 76	192 Ir 77	195 Pt 78	197 Au 79	201 Hg 80	204 Tl 81	207 Pb 82	209 Bi 83	210 Po 84	210 At 85	222 Rn 86
223 Fr 87	226 Ra 88	227 Ac 89															

Rare earths														
139 La 57	140 Ce 58	141 Pr 59	144 Nd 60	145 Pm 61	150 Sm 62	152 Eu 63	157 Gd 64	159 Tb 65	162.5 Dy 66	165 Ho 67	167 Er 68	169 Tm 69	173 Yb 70	175 Lu 71

Actinides														
227 Ac 89	232 Th 90	231 Pa 91	238 U 92	237 Np 93	242 Pu 94	243 Am 95	247 Cm 96	247 Bk 97	249 Cf 98	254 Es 99	253 Fm 100	255 Md 101	254 No 102	257 Lr 103

INDEX

Bold page numbers indicate tables.

Guildford College
Learning Resource Centre

Please return on or before the last date shown.
No further issues or renewals if any items are overdue.
"7 Day" loans are **NOT** renewable.

2 3 FEB 2012
- 6 NOV 2014